The Economics of Sustainable Urban Water Management: The Case of Beijing

Xiao Liang

Cover design: Chen Ru

CRC Press/Balkema is an imprint of the Taylor & Francis Group, an informa business

Published by:
CRC Press/Balkema
PO BOX 447, 2300 AK Leiden, The Netherlands
Email: Pub.NL@taylorandfrancis.com
www.crcpress.com - www.taylorandfrancis.co.uk - www.ba.balkema.nl

ISBN 978-0-415-69173-4

The Economics of Sustainable Urban Water Management: The Case of Beijing

Economische aspecten van duurzaam stedelijk waterbeheer in Beijing

Thesis

**to obtain the degree of Doctor from the
Erasmus University Rotterdam
by command of the
rector magnificus**

Professor dr H.G. Schmidt

and in accordance with the decision of the Doctorate Board

**The public defence shall be held on
October 5, 2011 at 16:00 hours
by**

**Xiao Liang
born at Hainan, China**

iss International Institute of Social Studies
Erasmus

Doctoral Committee

Promotor
Prof.dr. M.P. van Dijk

Other members
Prof.dr. J.W.A. Hafkamp
Prof.dr. A.H.J. Helmsing
Prof.dr. J.B. Opschoor

To My Parents and Husband

Contents

List of Tables and Figures

Tables

Figures

Acronyms

BNU	Beijing Normal University
CBA	Cost Benefits Analysis
DALY	Disability Adjusted Life Year
Gao	Gaobeidian plant
GIS	Geographic Information System
IRR	Internal Rate of Return
Jiu	Jiuxianqiao plant
NGO	Non-Governmental Organization
NPV	Net Present Value
O&M	Operation and Maintenance
Qing	Qingzhiyuan plant
SEPA	State Environmental Protection Agency
SWITCH	Sustainable Water Improves Tomorrow's Cities' Health
WHO	World Health Organization

Acknowledgements

Before I went to the Netherlands, I never thought I would pursue a PhD degree. The Master's programme at the Institute of Social Studies, The Netherlands (2004-2005) aroused my interest in continuing my academic studies. In 2006, I started the PhD programme at UNESCO-IHE Institute for Water Education to study water management. It was a challenge for me to begin the research in water management as my previous academic experience was economics. Fortunately, during the four years spent on my PhD journey, I conquered many difficulties on the path to completion of my PhD thesis. The PhD journey enhanced my ability to deal with problems independently, built my confidence, and extended my knowledge from economics to the environment. For family reasons, my PhD thesis was completed in three cities and countries: Delft (Netherlands), Metz (France) and Hong Kong (China). I would like to acknowledge the people who helped and supported me on this long journey.

I would like to express my appreciation and gratitude to my promoter Prof. Meine Pieter Van Dijk. Thanks to his great supervision, I learned how to carry out field work and data collection under the difficult situation of insufficient contacts and support. Under his supervision of the paper wiring, I learned that the attitude for scientific research should be critical and strict. I greatly appreciated his kindness and understanding as I split my time, for family reasons, between the Netherlands and France.

When I was doing research field work in China, I obtained assistance and support in gathering information, establishing contacts and data collection. Acknowledgements go to the Institute of Geographic Science & Natural Resources Research, Chinese Academy of Sciences, Beijing, China. I especially thank Prof. Cai Jianming, Drs Ji Wenhua, Zuo Jianbing, and Li Jiuyi. Their assistance enabled me to start my field work in Beijing. I would also like to thank the officials at Beijing Agro-Technical Extension Center for providing assistance on data collection of rainwater harvesting systems in

Beijing. Many thanks go to Prof. Zhou Jun at the R&D Center, Beijing Drainage Group for kindly sharing with me data on centralized wastewater reuse systems in Beijing. Thanks to Dr Zhai Jun (Chongqing University), Dr Bai Yongliang (China University of Geosciences, Wuhan), Ms Gan Guoqing (Wuhan), and Dr Zheng Xing (Germany), for their tremendous assistance during my field work in China.

I would like to express my gratitude to all my committee members for the time spent reading my thesis and their insightful comments to improve it.

Special appreciation goes to my former supervisor Dr Karel Jansen and the lecturers of my master's programme, Economics of Development at the Institute of Social Studies (ISS): Prof. Arjun Bedi, Prof. Mansoob Murshed and Dr Howard Nicholas. Thanks to their supervision and excellent teaching, I decided to continue my academic study.

I would like to thank my colleagues at UNESCO-IHE Institute for Water Education: Marco, Colin, Narrain, Zhu Xuan, Ye Qinghua, Wong Chee Loong, Yang Zhi, Hans, and the friends I made at the Institute of Social Studies: Wang Zhuoyu, Chia Dabao, Lu Caizhen, Tian Huifang, Wu Zhihui, for the scientific and non-scientific discussions, well wishes, support and assistance. Moreover, I thank my lovely friends in Delft and The Hague: Taoping, Han, Jun, Alwin, Ivette for their friendship. Especially, many thanks go to eight members of DGLZ (Da Guai Lu Zi) Hao, Huizhao, Xi, Wei, Hui, Zhuo, Bingyan, and Fanzhong for the wonderful trips, parties, skiing, caring and support.

I would like to thank my parents for being supportive all along. They gave me the freedom to choose my life although I am their only child. Thanks for their love and support. I am proud to be their daughter.

Last, special thanks goes to my dear husband, Huang Mingxin, for his encouragement when I met with difficulty, his unconditional love and support, and his tolerance of the bad temper of the 'PhD lady'. He is not only my husband but also my best friend.

Abstract

A rapidly growing urban population leads to the dramatic increase of water consumption in the world. The water resources available to the human being are limited. Meanwhile climate variability and environmental pollution decrease the quantity of water resources available for human use. It is a significant challenge to provide sufficient water to urban residents in a sustainable and effective way. Facing urban water crisis, researchers point out a paradigm shift in urban water management for sustainable water supply and services. This requires multi-disciplinary approaches, including technical improvements and economic evaluations. Advanced technology can contribute to the solution of problems physically, but it may not ensure sustainable operation of water systems. The obstacles to sustainable water supply and services often are from non-technical problems such as low cost recovery, lack of sound pricing systems and sustainable financing for increasing service coverage. The financial and economic factors could be a large barrier to the operation of water systems.

This research aims to use economics to assess water systems for sustainable urban water management. How to use economics on urban water systems and what contributions can economics bring to sustainable water management are the two main research questions in the thesis.

Since the existing systems are insufficient to achieve the objective of sustainable urban water management, many new systems are being proposed and implemented recently. There are two kinds of water systems: traditional or existing systems, and new or alternative systems. The alternative systems may be technologically feasible to increase water supply or save water consumption, but they may not be financially and economically feasible. Lack of financial and economic viability makes alternative systems less attractive than traditional systems. It is important to know whether the new systems can operate long term and whether the new systems are suitable alternatives to existing systems if one wants to promote sustainable urban water management.

The thesis carries out economic and financial analysis of traditional and alternative urban water systems. A comparative analysis between the traditional and alternative water systems is presented. Through the comparative analysis, the thesis shows whether the alternative system is an economically viable alternative to the traditional system. The case of Beijing is chosen for the study. The main technological measures of water saving in Beijing include wastewater reuse and rainwater harvesting. There are centralized and decentralized wastewater reuse systems. Centralized wastewater reuse systems represent the traditional systems while decentralized systems represent the alternative systems. Groundwater is the main and traditional water resource for agricultural irrigation, and rainwater harvesting is an alternative method to get more water.

The main economic method in the thesis is cost benefit analysis, which is an accepted method to evaluate the environmental projects. Additionally, the thesis employs the methods of linear programming and rough set analysis. In the cost benefit analysis, the concern of different stakeholders having different viewpoints is taken into consideration. Accordingly, an integrated financial and economic analysis is carried out, in which financial analysis is implemented from the point of view of individual participants, while the economic analysis is from the point of view of society. The financial analysis aims to judge whether the individual investor could afford the water system, and the economic analysis is to determine the contribution of the water system to the development of society.

The research shows that the alternative water systems are economically feasible while they are not financially feasible. However, the traditional water systems are both economically and financially feasible. Comparing the economic and financial feasibility between the traditional and alternative water systems, the traditional water systems are better than the alternative systems. It implies that the new water systems are not viable alternatives to the traditional water systems because the new systems are not financially feasible.

Through the case of Beijing, the thesis demonstrates how to use economics in managing urban water systems. This is the first integrated and quantitative analysis of the economic, environmental and social effects of new water systems. The economic, environmental and social effects are all determined by monetary values, which is rare in the existing literature.

The thesis shows that economics contributes to identifying the non-technical problems in water systems and can help decision makers to make choices that are consistent with the long-term well being of the community. Three practical contributions of the research are as follows. 1.Using economics to identify and quantify the effects of water treatment systems on economics, environment and society; 2.Using economics to discern the factors that significantly hinder long term plant operations; 3.Using economic tools to learn the advantages and disadvantages of different water systems from an economic perspective. The theoretical contribution of the research is that it proves the importance of considering the viewpoints of different stakeholders in the cost benefit analysis. Doing cost benefit analysis from different stakeholder perspectives can provide complete and accurate information for helping decision makers to choose the most suitable alternative.

Samenvatting

Economische aspecten van duurzaam stedelijk waterbeheer in Beijing

Door de snelle bevolkingsgroei in de steden neemt de waterconsumptie wereldwijd sterk toe. De bestaande watervoorraden zijn beperkt en door de veranderlijkheid van het klimaat en de milieuvervuiling neemt de hoeveelheid water die voor menselijk gebruik beschikbaar is af. Het is een grote uitdaging om stadsbewoners op een duurzame en effectieve manier van voldoende water te voorzien. Met het oog op een dreigend tekort aan water in de steden wijzen onderzoekers op een paradigmaverandering op het gebied van stedelijk waterbeheer om een duurzame watervoorziening en dienstverlening te waarborgen. Dit vereist een multidisciplinaire aanpak met onder andere technische verbeteringen en economische evaluaties. Geavanceerde technologie kan bijdragen aan de praktische en technische oplossing van problemen, maar is geen waarborg voor het duurzaam functioneren van watervoorzieningsystemen. Het zijn vaak niet-technische problemen, zoals een lage kostendekking, een gebrek aan goede prijssystemen en aan duurzame financiering van een steeds groter dienstverleningsgebied, die een duurzame watervoorziening en dienstverlening in de weg staan. Dergelijke financiële en economische factoren kunnen het functioneren van watervoorzieningsystemen ernstig belemmeren.

Dit onderzoek analyseert watervoorzieningsystemen voor duurzaam stedelijk waterbeheer vanuit een economisch perspectief. De twee belangrijkste onderzoeksvragen van dit proefschrift zijn hoe stedelijke watervoorzieningsystemen vanuit een economisch perspectief onderzocht kunnen worden en in hoeverre economische theorieën en methoden kunnen bijdragen aan duurzaam waterbeheer.

Omdat het doel van duurzaam stedelijk waterbeheer met de bestaande systemen niet bereikt kan worden, worden er de laatste tijd veel nieuwe systemen voorgesteld en ingevoerd. Er zijn twee soorten watervoorzieningsystemen: de traditionele of bestaande, en de nieuwe of alternatieve systemen. De alternatieve systemen zijn mogelijk in technologisch opzicht geschikt om de watervoorziening uit te breiden of water te besparen, maar ze zijn wellicht financieel en economisch niet haalbaar. Als alternatieve systemen financieel en economisch niet haalbaar zijn, zijn ze minder aantrekkelijk dan traditionele systemen. Als men duurzaam stedelijk waterbeheer wil bevorderen, is het belangrijk om te weten of de nieuwe systemen op de lange termijn goed kunnen functioneren en of ze een geschikt alternatief vormen voor bestaande systemen.

In dit onderzoek is een economische en financiële analyse van traditionele en alternatieve stedelijke watervoorzieningsystemen uitgevoerd. Dit proefschrift bevat een vergelijkende analyse van de traditionele en alternatieve watervoorzieningsystemen. Hieruit blijkt of het alternatieve systeem een economisch haalbaar alternatief is voor het traditionele systeem.

Het onderzoek heeft plaatsgevonden in Beijing. In deze stad zijn hergebruik van afvalwater en opslag van regenwater de voornaamste technologische maatregelen om water te besparen en op te slaan. Er bestaan gecentraliseerde en gedecentraliseerde systemen voor het hergebruik van afvalwater. De traditionele systemen zijn gecentraliseerd en de alternatieve systemen zijn gedecentraliseerd. Grondwater is de belangrijkste en traditionele bron voor de irrigatie van landbouwgrond, en de opslag van regenwater is een alternatieve manier om meer water te verkrijgen.

De belangrijkste economische onderzoeksmethode in dit proefschrift is kosten- batenanalyse, een gebruikelijke methode om projecten op milieugebied te evalueren. Daarnaast is gebruik gemaakt van lineair programmeren en rough set-analyse. In de kosten- batenanalyse wordt rekening gehouden met de belangen van verschillende belanghebbenden die uiteenlopende gezichtspunten hebben. Op deze manier is er een geïntegreerde financiële en economische analyse uitgevoerd. De financiële analyse gaat uit van het oogpunt van de individuele deelnemers, terwijl de economische analyse uitgaat van het maatschappelijk gezichtspunt. Het doel van de financiële analyse is om te beoordelen of het watervoorzieningsysteem betaalbaar is voor individuele investeerders, en met de economische analyse wordt de bijdrage van het watervoorzieningsysteem aan de ontwikkeling van de maatschappij bepaald.

Uit het onderzoek blijkt dat de alternatieve watervoorzieningsystemen wel economisch, maar niet financieel haalbaar zijn. De traditionele watervoorzieningsystemen zijn echter zowel economisch als financieel haalbaar. Uit de vergelijking van de economische en financiële haalbaarheid van de traditionele en alternatieve watervoorzieningsystemen blijkt dat de traditionele watervoorzieningsystemen beter zijn dan de alternatieve. Dit betekent dat de nieuwe watervoorzieningsystemen geen bruikbaar alternatief zijn voor de traditionele watervoorzieningsystemen omdat de nieuwe systemen financieel niet haalbaar zijn.

Dit proefschrift toont op basis van de situatie in Beijing aan hoe economische inzichten en onderzoeksmethoden toegepast kunnen worden op het beheer van stedelijke watervoorzieningsystemen. Dit is de eerste geïntegreerde en kwantitatieve analyse van de economische, milieu- en maatschappelijke effecten van nieuwe watervoorzieningsystemen. De economische, milieu- en maatschappelijke effecten worden allemaal herleid tot hun economische waarde, wat zelden voorkomt in de bestaande literatuur.

Dit proefschrift toont aan dat de economische wetenschap een bijdrage levert aan het vaststellen van de niet-technische problemen van watervoorzieningsystemen en beleidsmakers kan helpen keuzes te maken die op de lange termijn het welzijn van de gemeenschap dienen. Het onderzoek levert drie praktische bijdragen:

1. Het gebruik van economische inzichten om de economische, milieu- en maatschappelijke effecten van waterbehandelingssystemen vast te stellen en te kwantificeren.

2. Het gebruik van economische inzichten om de factoren aan te wijzen die het functioneren van watervoorzieningsinstallaties op de lange termijn ernstig hinderen.

3. Het gebruik van economische methoden om de voor- en nadelen van verschillende watervoorzieningsystemen te ontdekken vanuit economisch perspectief.

De theoretische bijdrage van dit onderzoek is dat het aantoont hoe belangrijk het is om de gezichtspunten van verschillende belanghebbenden in aanmerking te nemen in de kosten- batenanalyse. Het uitvoeren van een kosten- batenanalyse vanuit het gezichtspunt van verschillende belanghebbenden kan volledige en juiste informatie opleveren die beleidsmakers kan helpen om het geschiktste alternatief te kiezen.

1 Introduction

1.1 Problems in Cities

Global urban water utilization by people increased over 20 times in 100 years. In 1900, it was $200\times10^8\,m^3$; in 1950, it was $600\times10^8\,m^3$; in 1975, it was $1{,}500\times10^8\,m^3$; and in 2000, it was $4{,}400\times10^8\,m^3$ (Bao and Fang 2007). It is predicted that urban water utilization by people in the year of 2050 will equal total global water utilization in 2004 (Song et al. 2004).

Figure 1.1
Urban Population Growth

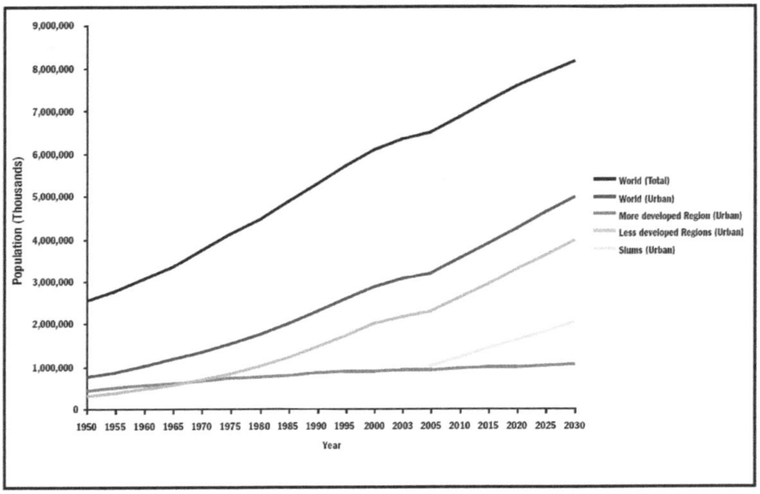

Source: Rees (2006)

A rapidly increasing urban population is an important factor causing the increase of water consumption in the world. Forty years ago, the urbanized population represented only 37 per cent of the total world population, but presently around 50 per cent of the world's population inhabits urban areas (Rees 2006). Figure 1.1 shows that the urban population is growing rapidly, especially in the developing regions. The urbanized population proportion in the developing regions may increase from 42 per cent to 57 per cent by the year of 2030 (Jenerette and Larsen 2006).

Rees (2006) thinks that the urban population may be more than 60 per cent of the total population by the year 2025. The number of megacities with more than five million residents is expected to increase globally from 46 to 61 between 2015 and 2030 with disproportionate increase in Asia and Africa (UN 2004). In China, approximately five cities have populations over ten million. Figure 1.2 indicates the growth of the urban population and urban water utilization in China, which shows that Chinese urban water consumption increases gradually while the urban population swells.

Figure 1.2
Growth of urban population and urban water consumption in China

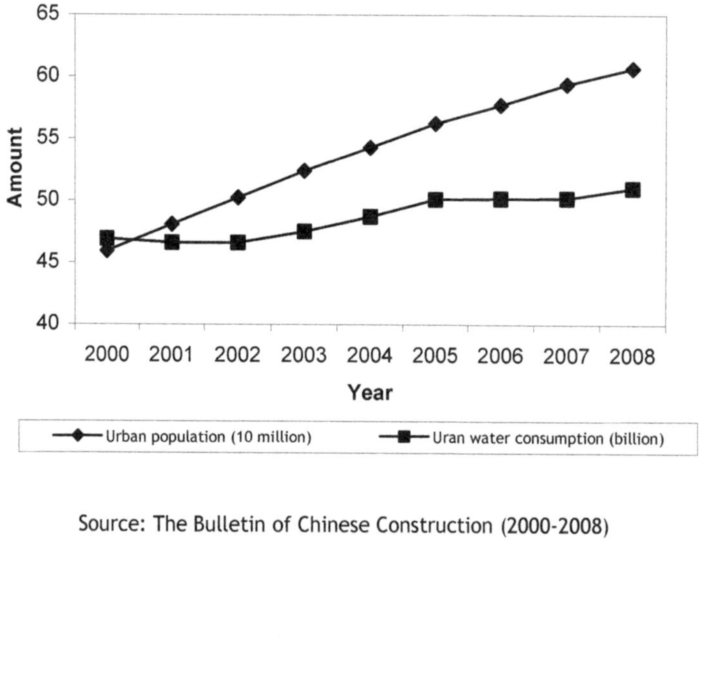

Source: The Bulletin of Chinese Construction (2000-2008)

Global water resources available to humans are limited. Only 2.5 per cent of the earth's 139×1016 m^3 of water is fresh water and less than one-third is available for human use (Postel et al. 1996). In China, renewable water resources equal only 2,205 m^3 per capita per year, which is one-fourth of the average world level (FAO, Water resource, Development and Management Service 2003). Per capita water availability in the 3-H basins of north China (Hai, Huai and Huang) is around 500 m^3 per year, which is well below the 1,000 m^3 per year standard for water stress (World Bank 2007). In Beijing, the total availability of water resources per capita per year is only 300 m^3, which is 1/8 of the national average and 1/32 of the world's average (Wang and Wang 2005). The limited water availability results in the shortage of water supply in cities.

Climate variability and environmental pollution lead to water scarcity in cities. Water crises affect not only arid areas but also some regions with normally plentiful water resources. For instance, some parts of Europe, have suffered successive droughts over the last few years, with the result that certain watercourses have dried up (Lazarova et al. 2001). The southwestern area of China, which used to have rich water resources, suffered a sudden drought in 2010 leading to severe water scarcity. Moreover, environmental pollution decreases the availability of clean water, increasing the pressure of urban water supply. Rapid industrialization around the cities leads to serious water pollution in the cities. Consequently, the clean water available to the urban residents diminishes. For example, due to pollution in many of the lakes that comprise the main domestic water source, many residents of the rural areas of Wuhan city cannot access clean water.

Rapid population growth, limited water availability, climate variability and environmental pollution together place significant pressure on urban water management, especially in arid areas. Urban population growth increases the demand for water, but water resources remain limited. Meanwhile climate variability and environmental pollution decrease the quantity of water resources available for human use.

1.2 Water Scarcity in Beijing

The thesis studies the context of Beijing, China because Beijing is a typical case of urban water scarcity. Increasing population, continual droughts and depletion of groundwater stocks pose a challenge to provide sufficient water to Beijing's residents in a sustainable and effective way.

The critical issue is that water consumption within Beijing is more than the water available to Beijing. Figure 1.3 illustrates that water consumption is almost one billion cubic metres more than the water available although the gap between water consumption and water availability decreases gradually.

Figure 1.3
Water availability and water consumption in Beijing

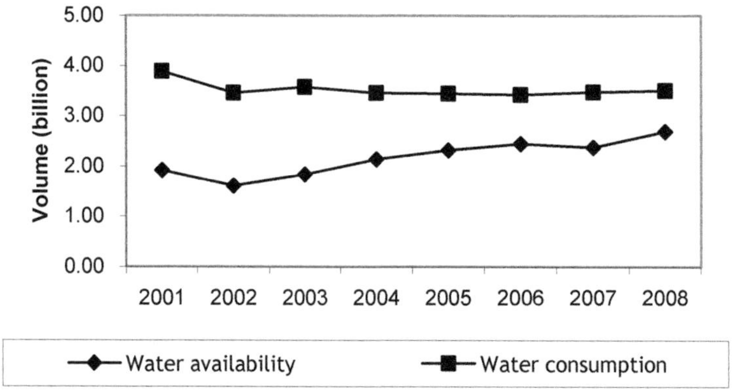

Source: Beijing Statistical Yearbook (2001-2008)

Water scarcity in Beijing is caused by two important factors: 1) Beijing has a large and increasing population; 2) Beijing is located in the arid area of China.

Beijing, the capital of China, is the second largest city after Shanghai and the political and financial centre of China. Important national governmental and political institutions, including the National People's Congress, are located in Beijing. Because of the dramatic economic development during the last 30 years, Beijing has been urbanizing rapidly, with an average annual increase of 2.48 per cent. Beijing is experiencing rapid economic development. Recently the average annual GDP growth rate was about nine per cent. Figure 1.4 shows that the population of Beijing increased, from 12.5 million in 1998 to 17 million people in 2008, an increase of nearly 1.5 times in ten years.

Figure 1.4
Population growth in Beijing from 1998 to 2008

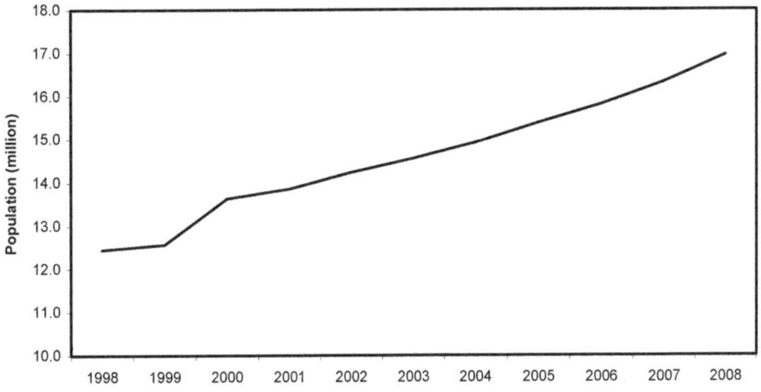

Source: Beijing Statistical Yearbook (1998-2008)

Figure 1.5
Precipitation rate in Beijing from 1986 to 2009

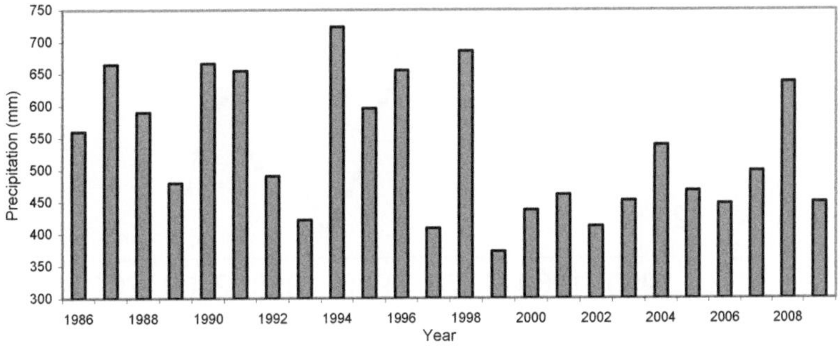

Source: Beijing Water Resources Bulletin (1986-2009)

Arid regions around the globe are most often associated with physical scarcity. Northern China, including Beijing is an area of physical water scarcity (Seckler et al. 1998). Beijing lies at the northern tip of the roughly triangular North China Plain, at an altitude of 20-60 metres above sea level. Beijing's climate is semi-humid monsoonal with a mean annual temperature of 10 to 12 centigrade. Mountains to the north, northwest and west shield the city from the encroaching desert steppes. The average altitude of the surrounding mountains is 1,000-1,500 metres. The Dongling Mountain located at the border of the Hebei province is the highest point in Beijing, with an altitude of 2,303 metres. Because of its geographical location, Beijing has low average rainfall. Beijing's average precipitation is 550 mm per year, 80 per cent of which falls between June and September (Beijing Water Authority 1986-2009).

Figure 1.5 illustrates the decrease in precipitation since 1999. The average precipitation between 1986 and 1998 was around 600 mm per year while the average precipitation between 1999 and 2009 was about 470 mm per year. In recent 10 years, precipitation decreased by 20 per cent, which led to less groundwater recharge.

Figure 1.6
Change in underground water level in Beijing (masl = metres above sea level)

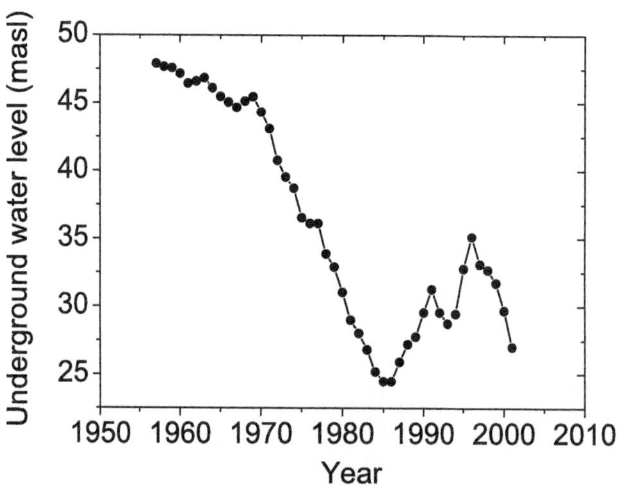

Source: Kaden (2007)

Groundwater is the main water source in Beijing, the city sources 70 per cent of total water supply from groundwater. However, overexploitation of groundwater due to increasing water demand and lower groundwater recharge both contribute to depletion of underground water stocks. Underground water levels in Beijing show significant decline since the mid-1950s (Figure 1.6). In rural areas of Beijing, the minimum depth of a well to access groundwater is around 80 metres deep while 20 years ago farmers could get groundwater from a well of only two metres depth. Figure 1.7 reflects the change in the depth of a well located in a village of the Huairou district of Beijing. It indicates that the depth of the well in 2007 was around 40 times deeper than in 1980. In some extreme cases, pumps no longer bring up groundwater and irrigation water has to come from 10 km away. The depletion of underground water stocks further complicates the difficulty of supplying sufficient water in Beijing.

Figure 1.7
Change in well depth in a village within the Huairou district of Beijing

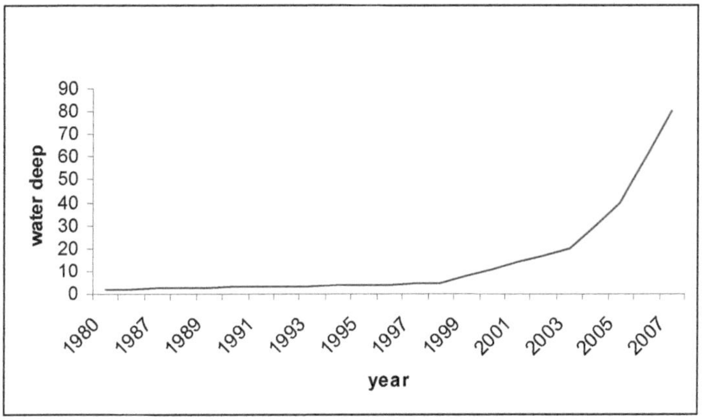

Source: Interviews with the farmers of the village

There are around 30,000 industries in Beijing, which accounts for 70 per cent of its GDP. Yet industrial sector water consumption accounts for only 15 per cent of total water consumption, domestic use and agricultural productions are the major water consumers accounting for 42 per cent and 34 per cent respectively according to data from 2008. The proportions of water consumption by agriculture, industry and domestic uses have changed in recent decades. Figure 1.8 shows the change in water consumption in agriculture, industry and for domestic use from 1989 to 2008. It shows that the water consumption of agriculture and industry are decreasing gradually while domestic water consumption is increasing steadily. Given the limited water supply and increasing population, more water is transferred from agricultural and industrial uses to domestic consumption. Accordingly the proportion of water consumption by agriculture, industry and domestic use has changed. Figure 1.8 shows that before 1998, agricultural water consumption was much greater than industrial and domestic water consumption, accounting for around 50 per cent of total consumption. Domestic consumption accounted for the smallest percentage of water usage. In 1998, domestic water consumption started to overtake industrial consumption although agricultural consumption is still the largest percentage. In 2006, domestic water consumption was greater than that of agriculture, becoming the largest water consumer. Domestic water consumption in Beijing rose from 1.4 billion m^3 in 1989 to 4.2 billion m^3 in 2008; agricultural water consumption decreased from 5.4 billion m^3 in 1989 to 3.4 billion m^3 in 2008; and industrial consumption declined from 3 billion m^3 to 1.5 billion m^3 during the same period (Figure 1.8).

Figure 1.8
Water consumption of agriculture, industry and domestic in Beijing

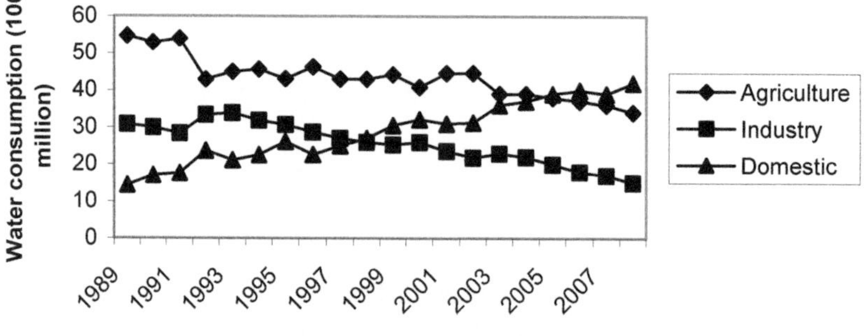

Source: Beijing Statistical Yearbook (1989-2008)

1.3 Chinese Urban Water Management

Because conventional urban water management has not adapted to the trends of urban development, new approaches to urban water management are proposed gradually. In China, many new technological measures are adopted in the cities to solve water scarcity. Additionally, the governmental structure of the water sector is changing and new water policies are issued.

1.3.1 Technological measures

Recently considerable research and experiments related to sustainable urban water management have been funded by the governments and carried out by water engineering scientists for the purpose of solving the water crisis in cities (Asano 2005; Chu et al. 2004; Wilderer and Schreff 2000; Zuo et al. 2010). All of the technological measures for developing alternative water resources, such as wastewater reuse, rainwater harvesting or transferring water from another source to the city, have been applied in Chinese cities (Bao and Fang 2007; Deng and Chen 2003; Jia et al. 2005; Zuo et al. 2010). The following sections offer details of wastewater reuse and rainwater harvesting.

The largest and most well known project for transporting water from remote areas to cities is being constructed in the east of China, called the 'South-to-North Water Diversion' project. This project aims to transfer water from the Yangzi River (South) to the areas of the 3-H basins (Hai River, Huai River and Huang River) (North). Beijing and Tianjin are the important beneficiaries of this project. The 'South-to-North Water Diversion Project' was first proposed in the 1950s and is expected to be completed in 2014 (Source: interviews with officials at the Beijing Water Authority). Up to 2010, approximately 60 billion Yuan has been invested in the project. The detail of the 'South-to-North Water Diversion' project refers to the official website of the project: (www.nsbd.gov.cn). The project will make a large contribution to the water supply of Beijing. The project is expected to transfer $7m^3$ per second of water if the project operates at full capacity. However, the financial feasibility of the project remains an open question. Since nine provinces and two provincial cities are involved in the project, there are complex issues concerning water rights, water allocation, water price and many others. Some NGOs (Non-Governmental Organizations) and researchers have criticized the project (Berkoff 2003; Liu 1998; Wang and Ma 1999).

Wastewater reuse

Wastewater reuse is the process of reclaiming grey water from industry and domestic sources and then reusing the water in industry as cooling water, domestically for toilet flushing and green irrigation, and in agriculture for irrigation. A conceptual cycling overview of urban water in Figure 1.9 illustrates the major pathways of water reuse and the potential use of reclaimed wastewater. The use of reclaimed water as an alternative water source can be perceived as an action of effective water management since fresh water is saved for other uses. It may prove sufficient flexibility to allow a water agency to respond to short-term needs as well as to increase long-term water supply reliability in urban areas (Asano 2001). Therefore wastewater reuse is a vital component of sustainable urban water management.

Figure 1.9
Roles of reclamation and reuse facilities in water recycling

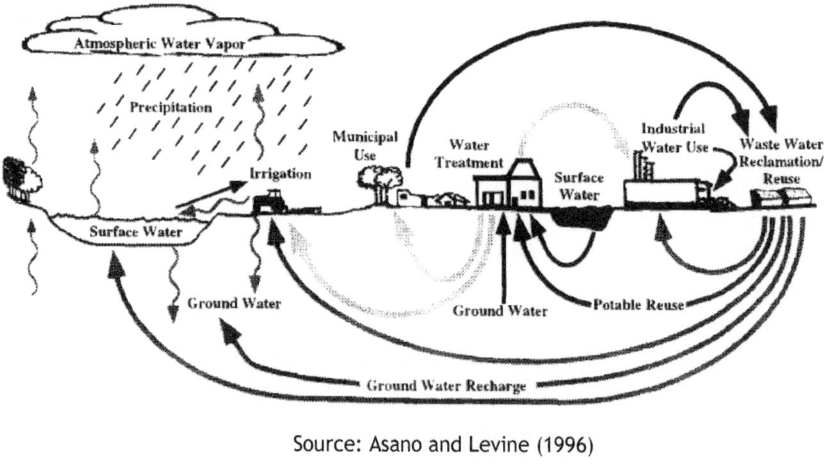

Source: Asano and Levine (1996)

Several Chinese cities, such as Beijing, Nanjing and Wuhan have implemented wastewater reuse. Beijing is the first city in China to force decentralized wastewater reuse systems and is the first city to build large-scale centralized wastewater reuse systems.

Wastewater reuse was first promoted in 1987 in Beijing through issuing the *Regulation of Building Decentralized Wastewater Reclamation Systems in Beijing*, which states that all hotels and residence areas with construction areas that exceed 20,000 m^2 and all other buildings with construction areas that exceed 30,000 m^2 must build wastewater reuse systems. Decentralized wastewater reuse systems collect and treat grey water and reuse reclaimed water on site. Generally decentralized wastewater reuse systems have small capacity and scales. Until 2002, in the central region of Beijing, there were more than 154 small wastewater reuse systems (Jia et al. 2005). Although decentralized water treatment systems have developed in some cities, they are still at the early stage in developing areas. To date, around 1,000 decentralized wastewater reuse systems have been constructed in Beijing (Zuo et al. 2010). The number of decentralized systems in Beijing is increasing and will continue to grow in the future.

In addition to decentralized wastewater reuse systems, large centralized wastewater reuse systems operate in Beijing. Centralized wastewater reuse systems collect grey water from different organizations and households, reclaim the collected grey water in a large plant, and then distribute the reclaimed water to users. Normally centralized wastewater reuse systems are large-scale operations. The first centralized wastewater reuse system in Beijing began operating in 2000. There are five centralized wastewater reuse systems in Beijing: the Gaobeidian wastewater reclamation plant (designed treatment scale: 470,000 m^3/day); the Jiuxianqiao wastewater reclamation plant (designed treatment scale: 60,000 m^3/day); the Fangzhuang plant (designed treatment scale: 10,000 m^3/day); the Wujiacun plant (designed treatment scale: 40,000 m^3/day) and the Qinghe plant (designed treatment scale: 80,000 m^3/day). Only two systems, the Gaobeidian and the Jiuxianqiao plants are in operation, the others are still under construction.

Although wastewater reuse in Beijing started 20 years ago, the quantity of reclaimed and reused water is still very small. In 2008, in Beijing, 0.6 billion cubic metres of water was reclaimed, accounting for 17 per cent of total water supply in Beijing (Beijing Statistic Yearbook 2008). Reclaimed water in Beijing is mostly used for domestic uses (toilet flushing, car washing and green irrigation), industry (cooling water) and water supplementation of rivers and lakes. Very little reclaimed water is used for agricultural irrigation. Another technological measure, rainwater harvesting, which aims to supplement water for agricultural irrigation, appears in the next section.

Rainwater harvesting

Rainwater harvesting is to induce, collect and store runoff from various sources for various purposes. Researchers in many places, such as in Sub-Saharan Africa, the Middle East and Asia have made efforts to develop rainwater harvesting for irrigation. In China, people have harvested rainwater for thousands of years, especially in the rural areas of the western north of China. Yet the most efficient use of rainwater for agricultural irrigation remains a subject of debate (Li et al. 2000; Mushtaq et al. 2007; Tian et al. 2002).

In urban areas, rainwater harvesting is rare because of the challenges of collecting and reusing water of sufficient quality. Many industrial zones are constructed around the city and domestic consumption creates various wastes so that air pollution is unavoidable in the city. This leads to poor quality rainwater. Normally the process of rainwater harvesting is easy and cheap, including the process of collecting, depositing and reusing. Given the poor rainwater quality, it is necessary to add water treatment processes, which could raise the cost of using rainwater in the city. Therefore rainwater harvesting and reuse in urban areas is more complex and costly than in rural areas.

According to Zuo et al. (2010), rainwater harvesting in Beijing has gone through three stages: 1) initial study and exploration (1981-1999); 2) intensive research (2000-2005); 3) implementation of projects driven by the government (2006-present). A large number of rainwater harvesting projects have begun in Beijing since 2006, of which some are located in the urban areas of Beijing and others are in rural areas. In the rainwater harvesting projects in the urban areas, the collected rainwater is usually used for toilet flushing, car washing and green irrigation. In rural areas, the rainwater is mostly used for agricultural irrigation.

Figure 1.10 illustrates the distribution of urban and rural areas of Beijing. Beijing occupies an area of 16,410 square kilometres, of which 8 per cent is urban areas and 92 per cent rural areas. According to the statistics, the number of rainwater harvesting projects in rural areas is greater than in the urban areas of Beijing (Zuo et al. 2010). Thus rainwater harvesting in Beijing is currently concentrated in rural areas for agricultural irrigation. Rainwater is the alternative water resource to replace ground water for supplementing the supply of irrigation water in rural areas of Beijing.

Figure 1.10
Distribution of Beijing's urban and rural areas

1.3.2 Water governance structures

Water governance in China

The water governance structure in China is a multi-layered hierarchy. The various agencies within the structure are divided by territory, function and rank. Figure 1.11 shows the hierarchy in terms of territory. The arrows lead from the order-giving body to order–receiving body.

Some agencies have specific functions. They are always headed by a ministry at state level and have corresponding agencies or bureaus at provincial, municipal, county and district levels. For example, there are the State Environmental Protection Agency (SEPA), provincial environmental protection agency and urban environmental protection bureaus. SEPA owns the provincial environmental protection agencies, which in turn owns the urban environmental protection bureaus. The head of SEPA appoints the heads of the provincial environmental protection agencies. The functional agencies exist within the different territorial ranks. So in different provinces or cities, the functional agencies concerned with water service and resource governance are almost the same.

Figure 1.11
Hierarchy of territorial rank

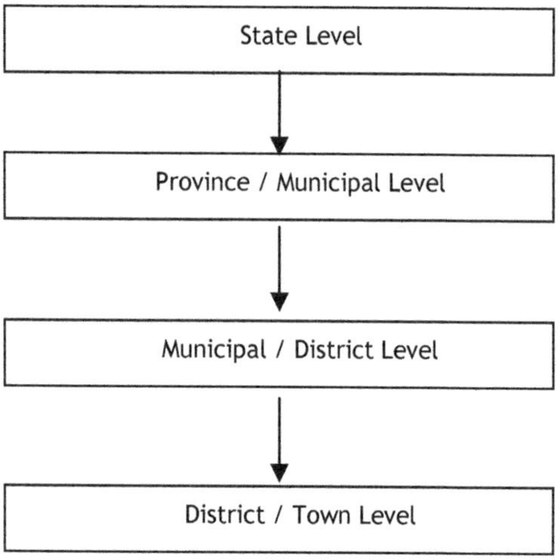

Every agency or bureau within the governance structure has a rank. Agencies and bureaus with the same rank cannot issue orders to each other. Only higher-ranking agencies can issue authoritative orders to lower-ranking agencies. For example, the Ministry of Construction has the same ranking as the Beijing municipal government. So the Beijing government could not issue a command to the Ministry of Construction. The communication should flow up and down level by level. It is rare that the provincial government issues commands to the town level skipping the municipal level.

The hierarchical governance structure creates a complex system of ministries, agencies and bureaus. It leads to an unclear policymaking process. Finding out how policies are made in China is much like tracing the movement of a single blood cell through the entire human body: the journey is time consuming and involves a network of organs and the specific route depends on the situation (Hou 2000). Figure 1.12 shows the old water governance structure of Beijing. It reflects the hierarchical order in the water sector and shows all kinds of agencies involved in water management in Beijing. In spite of the rigid structure, organization and cooperation between the agencies is poor.

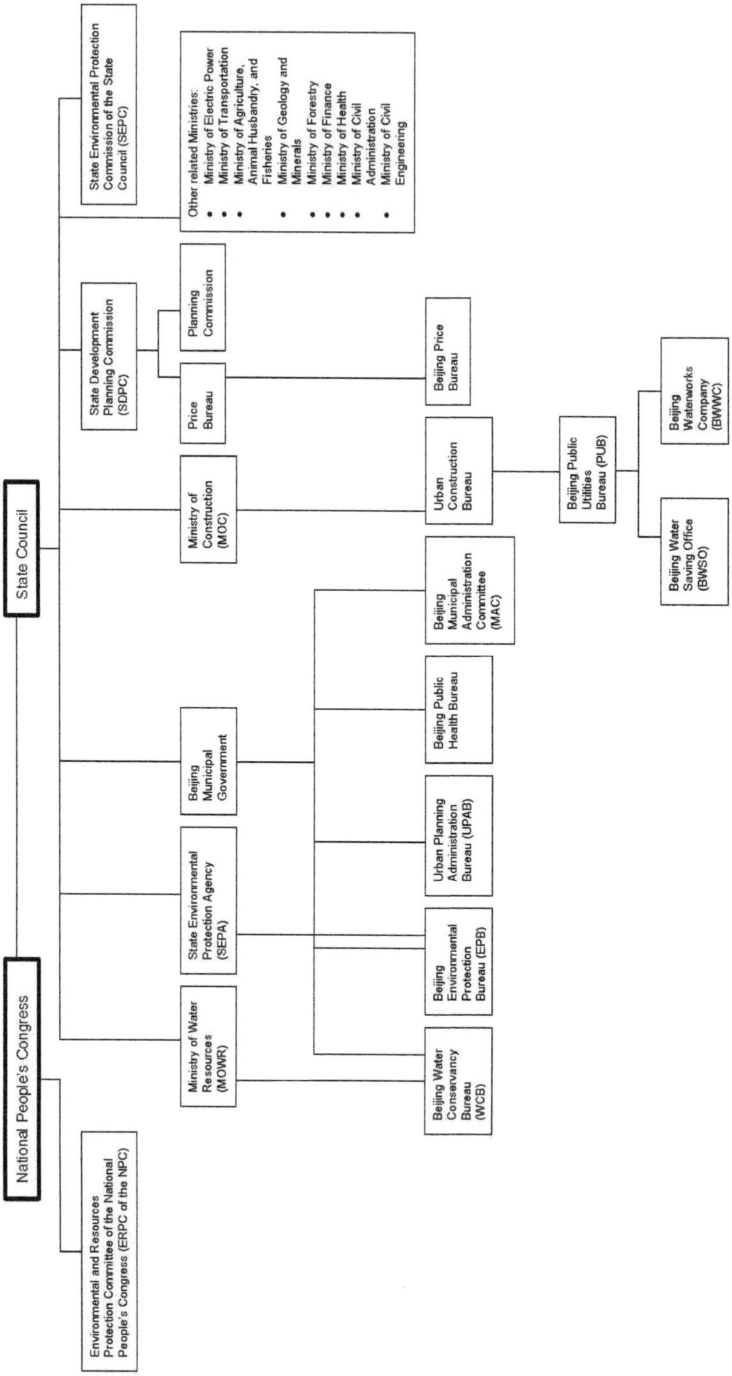

Figure 1.12

Organization of water relative governance in Beijing (Source: Hou, 2000)

Figure 1.13
New water-relative organizations of Beijing

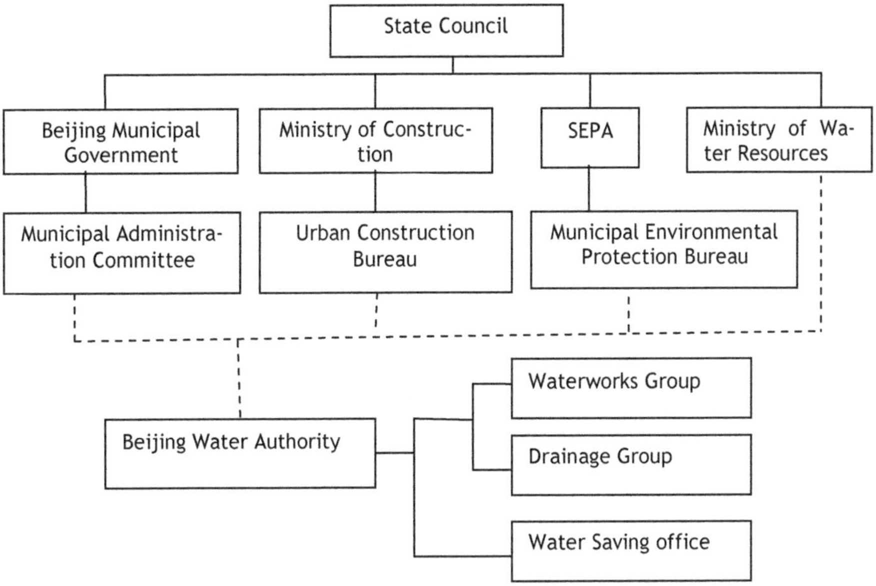

China is experiencing reform in water governance structures and institutions. The purpose of the reform is to reduce the number of agencies and bureaus involved in water management. The new water governance structure of Beijing, which could be an example for other Chinese cities and provinces, is shown in Figure 1.13. It reveals that in the new governmental structure there are four most relevant agencies: the Ministry of Water Resources, the State Environmental Protection Agency, the Ministry of Construction and the Beijing Municipal Government, involved in urban water management within Beijing. Beijing Water Authority is a new organization for water services and management, which was established in 2004 and is owned by the Beijing municipal government. It is in charge of water services and management at the different ranking territories. It is being established in many cities and provinces. There are two main functional organizations in the Beijing Water Authority: the Waterworks Group and the Drainage Group. The waterworks group is responsible for municipal water supply and the drainage group is responsible for municipal sewage treatment. Additionally the Water Saving Office (shown in Figure 1.13) is an important organization in the Beijing Water

Authority, in charge of promoting policy or project performance related to water saving.

According to the hierarchical structure, there will be provincial water authorities, urban water authorities and district/county water authorities. In Beijing, the Beijing Water Authority faces three layers of government from the municipal to the town levels (Table 1.1). Beijing has 16 districts and two counties, each district has sub-districts or, each county has towns.

Table 1.1
Three layers for water management of Beijing

Municipal Level	Beijing Water Authority	**Responsibilities:** Water resources, water supply and water pollution management etc. within the municipality. Coordinates among districts/counties.
District/ County Level	District/County Water Authority	**Responsibilities:** Water resources, water supply and water pollution management etc. within the district/county. Coordinates among sub-districts/towns.
Sub-district/ Town Level	Sub-district/Town Water Management Station	**Responsibilities:** Water resources, water supply and water pollution management etc. within the region.

Source: Pan (2006)

Water governance structures in Beijing to deal with water scarcity

Since the mid-1980s, continuous droughts and increasing population have worsened water scarcity in Beijing. Many projects concerned with water saving have been constructed and are promoted by the Beijing government. Because of the reform in the water governance structure, the number of organizations involved in the management of these newly constructed projects decreased. Meanwhile, to deal with water scarcity efficiently, many new organizations were established for the management of the projects related to water saving.

Figure 1.14
Gao plant main stakeholders

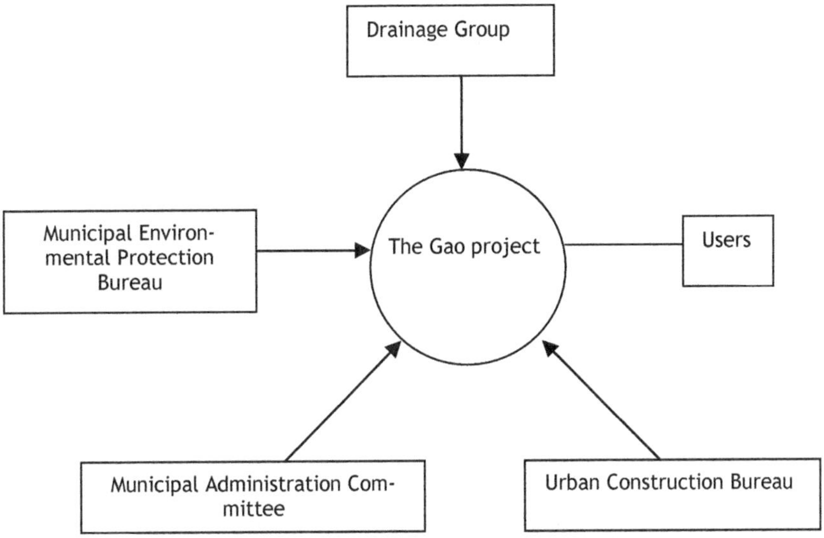

For the large water treatment plant, previously there were many organizations involved in management because it has a critical influence on society's health. This leads to problems in management of the large plants, which are often inefficient and sometimes ineffective. The reform of the water governance structure decreased the number of involved organizations, which may lead to improved management efficiency. The Gao Bei Dian wastewater reuse plant (Gao plant) in Beijing started in 2000 and is an example of the management of a large plant in the new governance structure. The Gao plant is owned by the Beijing Drainage Group, which belongs to the Beijing Water Authority. As shown in Figure 1.14, there are four main stakeholders in the Gao plant: the Drainage Group, the Municipal Environmental Protection Bureau, the Municipal Administration Committee and the Urban Construction Bureau. Under the 'old' structure of water governance in Beijing, there would have been eight organizations involved in the management of the Gao plant, while in the 'new' structure, the Beijing Drainage Group, as the owner of the Gao plant, is the key manager of the Gao plant. The functioning of the Gao plant is analysed in Chapter 3.

Figure 1.15
Qing plant main stakeholders

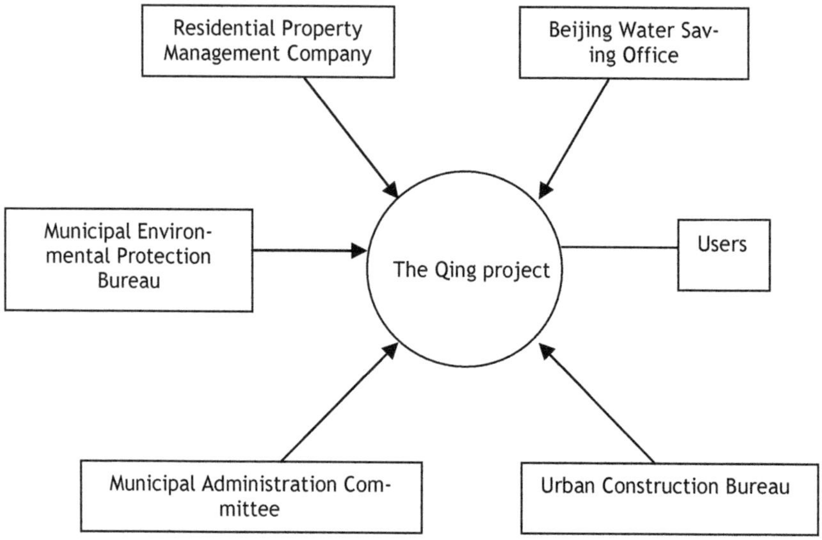

For decentralized treatment projects related to water saving, the Beijing Water Saving Office plays an important role. The Beijing Water Saving Office was established in 1981. Before the reform in water governance in Beijing, it was a small institute belonging to the Beijing Public Utilities Bureau. Previously the Water Saving Office was below the bureau level. However, after the reform in water governance structures in Beijing, the rank of the Beijing Water Saving Office increased to the bureau level. Due to serious water scarcity in Beijing, the promotion of water saving became more important. Consequently the responsibility of the Water Saving Office was enhanced. An example of the management of a decentralized water saving project is Qing plant (Figure 1.15). There are five main stakeholders involved in the management of the Qing plant. The Municipal Administration Committee and the Urban Construction Bureau (in charge of operation supervision), as well as the project manager and the Water Saving Office play important roles in managing the plant. In Chapter 2, an economic and financial analysis of the Qing project will assess its feasibility.

Figure 1.16
An plant main stakeholders

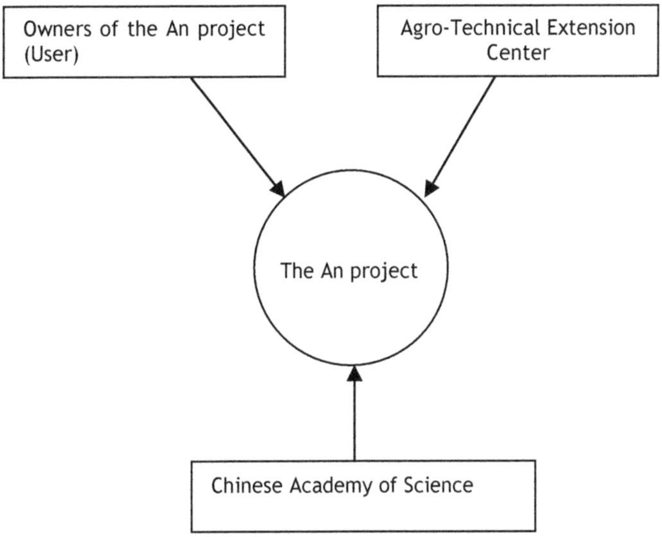

Many rainwater harvesting projects have been constructed in rural areas of Beijing to diminish water scarcity. The Beijing Water Saving Office supports some of these projects, and the Beijing Agro-Technical Extension Center supports others. Generally the owners of projects and the Beijing Agro-Technical Extension Center/ Beijing Water Saving Office are the main stakeholders of these projects. In some cases, academic institutions are involved in the management of the experiment. Figure 1.16 illustrates the main stakeholders of a rainwater harvesting project in rural areas of Beijing, the An plant. At the An plant, in addition to the owner and the Beijing Agro-Technical Extension Center, the Chinese Academy of Science is an important stakeholder. The detailed analysis of rainwater harvesting appears in Chapters 4, 5 and 6.

The Beijing government has promoted water saving in various sectors such as industry and agriculture. To extend water saving effectively, organizations in different sectors join in project management. For example, the Beijing Agro-Technical Extension Center is responsible for training farmers in new agricultural technologies and for technical support. The Agro-Technical Extension Center is not a professional organization in the water sector, but it joins the management of rainwater harvesting systems. Other academic institutions such as the Chinese Academy of Science are also involved in the management of these projects.

1.3.3 New policies

There are several crucial laws concerning water resource management in China. The *Water Law*, issued in 1988 and revised in 2002, aims to undertake the rational development, utilization, saving and protection of water resources. Additionally there are the *Water Pollution Prevention Law*, the *Water and Soil Conservation Law*, the *Environmental Protection Law* and other relevant laws and regulations. The *Water Pollution Prevention Law*, issued in 1984 and amended in 1996, plays an important role in preventing water pollution. The *Water and Soil Conservation Law* issued in 1991, emphasizes protection and conservation of water and soil recourses. The *Environmental Protection Law* issued in 1989 aims to protect and conserve national environmental resources.

In response to legal institutions established at the national level, the provincial and municipal governments are able to establish their own legal systems appropriate to local circumstance. For example, due to the rapid development of urbanization and industrialization, wastewater discharges into rivers, lakes and trenches without serious treatment. This means that a number of Chinese cities have experienced serious water pollution. The State issued the national level *Water Pollution Prevention Law* to help prevent water pollution in urban areas. Accordingly different regulations were issued at the regional level. For example, in Beijing, there are the *Quality Standard of Wastewater Discharge in Beijing* and the *Guan Ting Dam Water Conservation.*

As water scarcity is a serious problem in Chinese cities, many regulations concerning water saving were issued in some Chinese provinces or cities. For example, in Beijing, there is the *Regulation of Reward on Water*

Saving of Beijing and the *Regulation of Punishment on Water Waste of Beijing* issued in 1987 by the Beijing government. In 2000, a comprehensive regulation on constructing wastewater reuse systems was issued in Beijing. Standards for wastewater reuse were fixed, which include wastewater source standards, wastewater reclamation technique standards and reclaimed water quality standards. These new regulations in Beijing effectively promote and manage water saving projects and alternative approaches to solving water scarcity. Currently there is no law at the national level concerned with water saving.

There are still problems with policymaking and implementation. First, because the existing governance structure has not changed entirely, there are still conflicts between ministries involved in water management. For example, the Ministry of Water Resources oversees water resource management in China, and the State Environmental Protection Administration is in charge of water quality control. The duties of two ministries overlap at the area of water pollution. The undefined and overlapping responsibilities in water policy implementation between the two ministries result in unnecessary duplication of data collection and incoherent implementation of water quality control policy (Feng et al. 2006; Lee 2006). Second, definitions in many policies are ambiguous, especially in the area of agricultural irrigation. For example, the *Water Law* states that 'water belongs to the state; and people are encouraged to explore and utilize water resources'. It does not specify what kinds of organizations or individuals have the right to utilize water resources and obtain benefits from water exploration, resulting in unclear definition of water rights. This prevents water saving, efficient use of water and water trade between different regions (Lee 2006).

1.4 Economics for Sustainable Urban Water Management

Technological measures and water policies encourage a number of new water treatment systems to be constructed in cities. Advanced technology can contribute to the solution of problems physically, but it may not ensure sustainable operation of water treatment systems.

Problems of sustainable urban water management are often a series of non-technical factors. Most water cycle systems could not reach operational sustainability because of low cost recovery, lack of sound pricing systems and sustainable financing for increasing service coverage, particularly for the poor. For example, in China, the current pricing system does not reflect the real cost of water services in the city, which causes low cost recovery on water services. But local governments are still reluctant to increase the water price because they worry that increasing the price would have a negative impact on the local economy (Lee 2006). The financial and economic factors can become a big obstacle to sustainable performance of water systems.

Decision makers and scientists from different fields pursue sustainable urban water management as a general objective. Sustainability refers to the capability to maintain or improve standards of well-being over generations (Pearce 1989). From an engineering perspective, sustainable water management means water supply could effectively satisfy water demand continually and effectively; while from an economic perspective, sustainability of water management means effective allocation of available resource to reach maximum utility in the water sector. Different definitions for sustainable urban water management from different engineering and economic perspectives reflect that the sustainable urban water management requires multi-disciplinary studies, such as technological, institutional and economic studies. Falkenmark and Lundqvist (1998) think that water problems are mostly caused by competition among water uses and political, technological and economic barriers that limit access to water resources. Hence an economic analysis of water systems is a necessary part of sustainable urban water management.

Although investments in technological measures for water saving exists and many engineers are working on these measures, issues of economical, environmental and social feasibility of these new systems are less often discussed in the literature (Jia et al. 2005; Zuo et al. 2010). In terms of content, existing literature concerned with economic study of water cycles could be classified by five issues: 1) cost studies (Abeysuriya et al. 2005; Gratziou et al. 2005; Maurer et al. 2006); 2) benefit analysis (Birol et al. 2005; Psychoudakis et al. 2005); 3) tariff structure and price analysis (Kim 1995; Renzetti 1999; Renzetti and Kushner 2000; Rogers et al. 2002; Seppala and Katko 2003; Singh et al. 2005); 4) water demand analysis (Arbues et al. 2003) and 5) financing issues (Almagro 2005; Nigam and Rasheed 1998). The research focuses on the issues of cost and benefits studies in urban water treatment systems. Because it is very difficult and complex to calculate the economic impact caused by the water plants quantitatively, most studies in the literature merely describe the cost or benefits of water treatment systems qualitatively or make a simple cost calculation for the water systems (Abeysuriya et al. 2005; Braden and Van Ierland 1999).

1.5 Research Objectives

The research aims to use economics to study water systems for sustainable urban water management quantitatively. How to use economics to study urban water systems and what contributions can economics make to sustainable water management are the two main research questions in this research.

Since existing water systems are inadequate to achieve the objective of sustainable urban water management, many new systems are being proposed and implemented recently. There are two kinds of water systems made up of urban water management: traditional water systems namely existing systems, and alternative water systems namely new systems. The alternative systems may be technologically feasible to increase water supply or save water consumption, but they may not be financially and economically feasible in the performance. If the alternative water systems are not more financially and economically feasible than the traditional systems, this may hinder sustainable operation of alternative systems. Whether the new systems could operate in the end and whether the new systems are suitable alternatives to existing systems are important to know if one wants to promote sustainable urban water management.

The thesis narrows down the research to an economic study of the traditional and alternative urban water systems. There is a comparative analysis between traditional and alternative water systems in the research. The comparative analysis explores whether the alternative system is an economically viable alternative to the traditional system.

The previous sections state that the main technological measures of water saving in Beijing include wastewater reuse and rainwater harvesting. About wastewater reuse plants, there are centralized treatment systems and decentralized systems. Centralized wastewater reuse systems represent traditional systems while decentralized wastewater reuse systems represent alternative systems. Rainwater harvesting is an alternative method to get more water resources for agricultural irrigation. Usually groundwater is the primary traditional water resource for agricultural irrigation.

Based on interviews with officials, Table 1.2 shows some statistical data about decentralized wastewater reuse and rainwater harvesting, which shows that there are around 2,000 constructed decentralized wastewater reuse plants in Beijing. The percentage of utilization of these constructed wastewater reuse plants is only around 30 per cent although wastewater reuse has been in use in Beijing for almost 20 years. Around 700 rainwater harvesting plants have been built in Beijing. The construction of rainwater harvesting plants in Beijing started in 2006. Similar to decentralized wastewater reuse systems, utilization of constructed rainwater harvesting plants is very low, at a rate of only 20 per cent. Performance of new constructed water systems is not as expected. Given the low rate of utilization of new systems, the objective of water saving is difficult to achieve. This research investigates the reasons behind the low rate of utilization of new water systems.

Table 1.2
Project data on water saving in Beijing

Projects	Number of constructed projects	Percentage of utilization
Wastewater reuse	2,000	30%
Rainwater harvesting	625	20%

Source: Jia et al. (2005), Wang et al. (2007) and interviews with officials.

The thesis conducted an economic study of centralized and decentralized wastewater reuse systems, rainwater harvesting systems and groundwater systems in Beijing. Additionally there is a comparison between centralized and decentralized wastewater reuse systems, and between groundwater and rainwater use.

The current research is part of The SWITCH (Sustainable Water Improves Tomorrow's Cities' Health) programme, supported by the European Union. The SWITCH programme seeks a paradigm shift in urban water management. It is expected to generate new efficiencies from integration of actions across the urban water cycle in order to improve the quality of life in cities. The strategy of SWITCH is development, application and demonstration of a range of tested scientific, technical and socioeconomic solutions that contribute to the achievement of sustainable and effective urban water management schemes. This study emphasizes the financial and economic analysis of the urban water management system.

1.6 Research Methodology

The main economic method used in this research is cost benefits analysis (CBA), which is a largely accepted method to evaluate environmental projects. Although various economic methods have been employed to evaluate water management such as life cycle analysis, multi-criteria analysis, cost effectiveness study, contingent valuation methods and multiple goal programming, the CBA method is most suitable method for this study (Ashley et al. 1999; Hauger et al. 2002). Based on the theory of welfare economics, the CBA method can help to illuminate the trade-off involved in making different kinds of investments (Arrow et al. 1996). It can inform decision makers about how scarce resources can be put to the greatest social good.

In addition to CBA, the methods of linear programming and rough set analysis are used in Chapters 5 and 6 respectively. Linear programming is a mathematical method of determining a way to achieve the best outcome in a given model of linear relationship. Through determination of linear programming, it is possible to discover optimal solutions for reaching maximum net profits of an environmental activity. Linear programming is used extensively in economic analysis (Becker 1994; Berbel and Gomez-Limon 2000).

Rough set analysis is an artificial intelligence method of data mining. In social science research, information collected is mostly qualitative data or part qualitative and part quantitative data. The rough set analysis could synthesize the approximation of concepts from the data and then classify the information (Pawlak 1982; Slowinski 1991). It is a suitable method to deal with the mixture of qualitative and quantitative data in social studies. In this study, the sample is small and the information is a mixture of qualitative and quantitative data. Using the method of rough set analysis helps us to identify critical factors affecting the success or failure of an environmental activity.

1.6.1 Evaluation framework

To make a valuable contribution to decision making, it is important to encompass the full range of private and public sector concerns (Campbell and Brown 2005). The research takes into account the fact that different decision makers with different points of view may have different judgments about the same event. One decision maker regards one effect as beneficial, but it can imply higher costs to the other one. For example, taxes are treated as costs from a private perspective while in the public case they are not treated as costs. Accordingly the concerns of the private and public sectors in water management are different.

In this research, project managers and the government represent the private sector and public sector, separately, who are the two main stakeholders in water management. Project managers and the government should have different points of view on water management. From the perspective of project managers, the net profit from an investment is more important. Whether a water system makes money or not determines the incentive of project managers to operate the system. The government is concerned mostly with whether a water system has positive influence on society. That a water system has positive influences on society benefits sustainable urban water management.

In this study, two parts of an analysis are carried out separately from the angle of private sector namely project managers and from the angle of public sector namely government. As shown in Figure 1.17, the first part is financial analysis, which takes the point of view of project manager, and the second part is economic analysis, which takes the point of view of government. The purpose of financial analysis is to judge how much the individual participant could live with the project, while the economic analysis aims to determine the contribution of a proposed project to the development of the total economy (Gittinger 1982). Both financial analysis and economic analysis are complementary in the study.

The integrated financial and economic analysis can systematically and completely assess a water system.

Figure 1.17
Evaluation framework

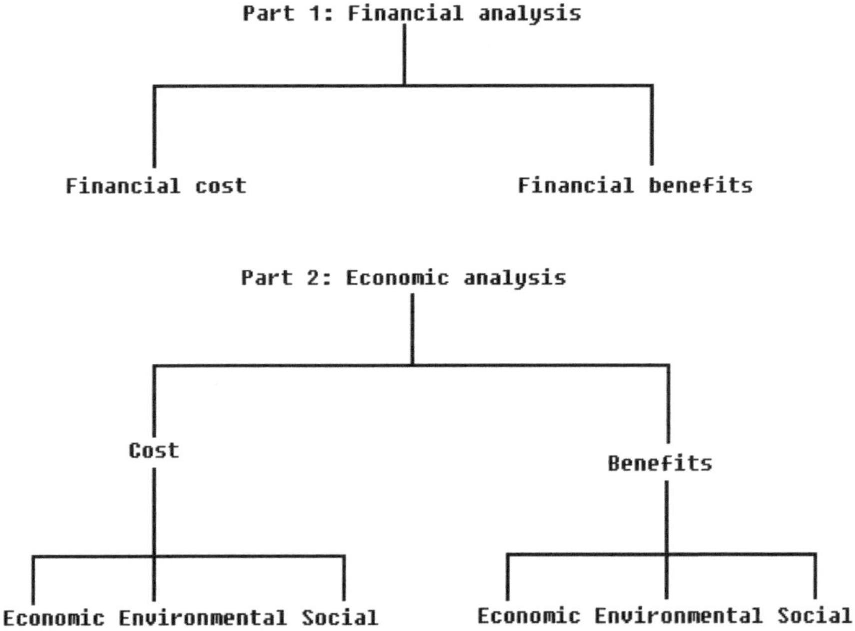

Since CBA method is used for the study, the values of cost and benefits of a water system are calculated quantitatively. Through calculating the present values of cost and benefits, the difference between benefits and cost could be obtained. The difference between benefits and cost determines whether the project is financially or economically efficient or not. Three different dimensions in financial and economic analysis are considered when using cost benefit analysis.

First, the cost and benefits considered in financial analysis and economic analysis are different. In financial analysis (shown in Figure 1.17), financial cost and benefits are evaluated. The contents of financial cost and benefits are identified from the private perspective. In economic analysis (shown in Figure 1.17), major economic, environmental and social cost and benefits are all quantified. From public perspective, a water system can cause an external influence on environment and society, so environmental and social effects are taken into consideration in the economic analysis. Traditional economic analysis of water systems usually ignores this aspect (Baumann et al. 1998).

Second, transfer payments such as subsidies and income from water charges are not considered in the economic analysis because they do not consume or create any new value for society (Dahmen 2000). But they may be considered in the financial analysis.

Third, the values used for determination of cost and benefits are different. In financial analysis, the market value is used directly for the value determination of financial cost and benefits. However, in economic analysis, monetary values of cost and benefits of economic, environmental and social effects are required in the assessment and can be calculated through indirect valuation method.

1.6.2 About the data

Interviews with the stakeholders of the water system were the main method of obtaining data from the plants studied. For example, the stakeholders of a decentralized rainwater harvesting system (shown in Figure 1.16) include the owner of the project namely the project managers, the Agro-Technical Extension Center and the Chinese Academy of Science. For this water system, most of the project's data was obtained through interviews with the owner of the project, and other data was obtained through interviews with the officials from Agro-Technical Extension Center and the involved researchers from the Chinese Academy of Science. Meanwhile the data obtained from one stakeholder required validation with other stakeholders to make sure the data was appropriate for the evaluation.

In addition to the interviews, other approaches were used to obtain the data. Some research data derive from existing literature such as internal reports of the Beijing Water Authority, the Agro-Technical Extension Center, and published articles.

1.7 Economic Theory

As mentioned in the methodology, cost benefit analysis is the main method used in the thesis. The ground principle of CBA is welfare economics, of which the essence is efficient allocation of scarce resources to maximize welfare. Efficient allocation implies that the possibility of relocating resources to achieve an increase in the net value of output produced by these resources (Hanley and Spash 1993). There are two main criteria to justify whether a project or policy could improve welfare. In classical welfare economics, welfare improvement is defined by the Pareto Optimality, which states that if nobody is made worse off and at least one individual believes he/she is better off, the project causes welfare improvement. The criticism of the Pareto Optimality is that most projects make some people better off and some people worse off simultaneously. It is hard to find a project that does not harm some people even if it benefits some people (Pearce and Nash 1981).

Modern welfare economics are based on the Kaldor-Hicks principle of potential compensation. In modern welfare economics, the criterion of welfare improvement is that if the gainers from an action could compensate the losers, the action is an improvement regardless of whether compensation is actually paid. In modern welfare economics, both individuals and society as a whole have a common goal of maximizing wellbeing and any project that relocates resources could make some affected people of society better off and others worse. All those made better off increase their own and society's wellbeing, while those made worse off suffer some burden and society also suffers this loss (Ganderton 2005). The hypothesis is that the people made worse off could receive compensation from the people made better off. The compensation in the Kaldor-Hicks principle is to be conceivable, which does not have to actually pay. The principle of potential compensation is based on consumer welfare theory and producer welfare theory, of which the details refer to Hanley and Spash (1993).

In terms of the principle of potential compensation, the benefits and cost generated by a project and the net benefits between benefits and cost can be measured. We can express net benefits as the difference between benefits and cost, or as the ratio of benefits to cost (Ganderton 2005). The effects on the gainers are regarded as benefits and effects on the loser are regarded as cost. According to modern welfare economic theory, if the discounted benefits caused by a project exceed the discounted cost, the project is regarded to lead to improve social welfare. This provides a benchmark for measuring the effects and performance of a project. The CBA method could help to reach efficient allocation of scarce resources in various sectors through measuring the effects caused by projects (Mishan 1988).

The definition and measurement of cost and benefits are essential parts in the CBA method. The concept of 'full cost' is pointed out, and is supposed to cover all costs pertained to water management. In the opinion of Rogers et al. (1998), the full cost of a water system consists of the capital cost, operation and maintenance cost, opportunity cost, economic externalities and environmental externalities. However the calculation of all the cost components is always difficult (Tsagarakis 2005). A well known problem is that environmental cost resulting from the project cannot always be valued based on existing market prices (Braden and Van Ierland 1999). Furthermore, benefits theoretically include all changes in resource use and services level. Estimating benefits is a complicated matter mainly because it is difficult to decide which benefits to include in the analysis and some benefits are intangible and difficult, if not impossible, to convert into monetary terms (Hauger et al. 2002). Few studies do a quantitative analysis of the social economic benefits (Birol et al. 2005; Psychoudakis et al. 2005). The measurement of cost and benefits should use 'true' economic values, which reflect the value of each resource (Ganderton 2005). Under restricted circumstances that the resources are not traded items and there are no large distortions in market prices of resources, market price could represent economic values. For some resources such as air, there are no market prices to reference. We can determine the values through indirect methods such as measuring value by the opportunity cost or value in alternative use method (Ganderton 2005).

In the CBA method, if benefits and cost stretch over time, present values of cost and benefits occurring in different periods are required. Net Present Value (NPV) is the difference between the present value of benefits and the present value of cost. When faced with a series of projects and a limited budget, choose projects in NPV order, with the highest NPV project selected first. In addition to NPV, there is another criterion to choose the project with high effectiveness: the internal rate of return (IRR). The IRR is extensively used for the investment appraisal of a project. It reflects the return on investment during the project operational period. IRR is the interest rate that equals the present value of benefits to the present value of costs. This means at the rate of IRR, the net present value of an investment equals zero. Given a series of projects and the same evaluation period, the project with highest IRR is selected first. The restriction of IRR method is that, under the circumstance of cost being much larger than benefits, the IRR cannot be obtained. This appears in the economic analysis of decentralized wastewater reuse systems in Chapter 2, in which the cost is much larger than the benefits. Hence NPV is more appropriate in this research because IRR cannot be calculated.

The discount rate used is another important issue in the CBA method. It is the rate at which future benefits and costs are discounted to present value (Prest and Turvey 1968). The discount rate is based on how individuals trade off current consumption for future consumption (Arrow et al. 1996). People use different discount rates for different kinds of things (Gintis 2000). For the evaluation of environmental resources, it is better to use the social discount rate (Arrow et al. 1996; Brent 1996). According to the publication *Chinese Economic Evaluation Parameters on Construction* (2006), the social discount rate used for cost benefit studies in China is 8 per cent including the inflation rate. Inflation rates in China for the years 2007 and 2008 are 4.8 per cent and 5.9 per cent, respectively, and the opportunity cost of capital is around 3 per cent (China Statistical Yearbook 2007, 2008).

1.8 Research Outline

Two major technological measures apply to water saving in Beijing: wastewater reuse and rainwater harvesting. The main contents of the thesis consist of two parts: the study of wastewater reuse (Chapters 2 and 3) and the analysis of rainwater harvesting (Chapters 4, 5 and 6).

The wastewater reuse projects in Beijing are mostly concentrated in the urban areas, and can be divided into large scale (centralized systems) and small scale (decentralized systems). Chapter 2 carries out the financial and economic analysis of the decentralized wastewater reuse systems in Beijing through the method of CBA. In Chapter 3, the financial and economic analysis of centralized wastewater reuse systems is carried out and then an economic comparison between the centralized wastewater reuse systems and decentralized systems is implemented. Chapters 2 and 3 study the financial and economic feasibility and cost recovery of the wastewater reuse systems in Beijing, and compare two kinds of water treatment systems: centralized and decentralized systems from a financial and economic perspective.

The rainwater harvesting systems are located in both urban and rural areas of Beijing. As we had no access to data on rainwater harvesting systems in urban areas, there is no analysis of the rainwater harvesting systems in urban areas in the thesis, which is the limitation of the thesis. The thesis focuses on rainwater harvesting in rural areas of Beijing. In Chapter 4, financial and economic analysis of rainwater harvesting in rural areas of Beijing is carried out, using the method of cost benefit analysis. In Chapter 5, a further discussion concerned with the charge of groundwater and the demand of rainwater is carried out using linear programming method. Chapter 6 tries to find the decisive factors affecting operation of rainwater harvesting in rural areas of Beijing through the rough set analysis. Chapters 4, 5 and 6 make a study of the financial and economic feasibility of rainwater harvesting systems in rural areas of Beijing, and discuss water price, cost recovery and decisive factors for the operation of rainwater harvesting systems. Chapter 7 draws the conclusions of the thesis.

2 Financial and Economic Analysis of Decentralized Wastewater Reuse Systems[1]

2.1 Introduction

Decentralized wastewater reuse system means that wastewater is collected and transferred to a plant on site, and then the reclaimed water is reused close to the plant. The main uses for reclaimed water are in urban domestic applications, such as toilet flushing, green irrigation or industrial cooling water. Started in the 1950s, decentralized wastewater reuse for domestic application in cities has been well developed in Japan thanks to promotion and investment by the government (Yamagata et al. 2003). Similarly, the government promoted and subsidized the decentralized wastewater reuse system in Beijing in 1980s. The Beijing government subsidizes investment of buildings and equipment. As mentioned in section 1.3.3 of Chapter 1, the Beijing government issued *The Regulation of Building Decentralized Wastewater Reclamation Systems in Beijing*, requiring large organizations to build their own wastewater reuse systems. Therefore a substantial number of decentralized wastewater reuse systems have appeared urban areas of Beijing.

The performance of decentralized wastewater reuse systems in Beijing is not as good as expected. The average utilization of wastewater reuse systems is only around 30 per cent, as indicated in Table 1.2 of Chapter 1. The operation of some small wastewater reuse systems has been suspended.

The existing technology for wastewater reuse has developed to the point where it is technically feasible to produce water of any quality

(Asano 2005). Small wastewater reuse systems are now capable of producing reclaimed water in a reliable way. Thus the poor performance of the decentralized wastewater reuse systems in Beijing may be caused by economic problems. To become competitive, a system must achieve both physical and economic efficiency. It is necessary to determine the economic feasibility of the decentralized wastewater reuse systems in Beijing.

Several researchers have carried out studies of the financial and economic feasibility of wastewater reuse systems. These papers either try to prove that the technologies are economically feasible and worth further development, or they seek to find the relationship between the scale of treatment plant and the cost of running it (Friedler and Hadari 2006; Maurer 2009; Nurizzo et al. 2001; Tsagarakis et al. 2000; Yamagata et al. 2003). It is rare that one study evaluates both financial and economic feasibility. Moreover, generally, only internal costs such as initial investments and operation and maintenance costs are taken into consideration. Few papers try to quantify the environmental and social effects (Genius et al. 2005; Tziakis et al. 2008).

This chapter aims to create an integrated financial and economic feasibility analysis of decentralized wastewater reuse systems in Beijing. The main stakeholders of decentralized wastewater reuse systems are the project manager and government. Different decision makers may have different points of view on the same event. Based on that, this chapter carries out the study from the point of view of project manager and from the perspective of government separately. The financial analysis takes the viewpoint of individual participants, namely the project manager, while the economic analysis is from the perspective of government.

Since cost benefit analysis is the evaluation instrument for the analysis in this study, the present values of benefits and cost are calculated for comparative analysis. As illustrated in Figure 1.17, the financial analysis encompasses an evaluation of the financial cost and benefits, assessing the financial performance of the investments. In the economic analysis, the major economic, environmental and social effects are selected and quantified.

Section 2.2 introduces the case studies, which are representative cases for Beijing. Sections 2.3 and 2.4 present separately, how to carry out both financial and economic analyses. Section 2.5 provides the results of the analyses and discussion of the results. The conclusions are provided in the last section.

2.2 About the Decentralized Wastewater Reuse Plants

Two cases, the Qingzhiyuan (Qing) plant and the Beijing Normal University (BNU) plant, have been chosen for the analysis. These two plants are representative cases in Beijing for three reasons. First, the Qing and the BNU plants are both located in the city centre of Beijing. Currently the decentralized wastewater reuse systems are all in the urban areas. Second, the two plants process grey water reclamation and reuse for toilet flushing and green land irrigation, which is similar to other wastewater reuse plants. Third, the technology used at the Qing and the BNU plants is similar to other wastewater reuse plants.

The Qing plant is located in a residential area and serves approximately 2,500 people. The BNU plant is located on a university campus and serves approximately 30,000 people. The treatment capacity of the Qing plant is around 65 m^3 per day and the capacity of the BNU plant is 400 m^3 per day. As the wastewater treatment technology of the Qing plant is similar to that of the BNU plant, it is possible to make a direct comparison between these two plants. All data for the estimation are collected through interviews with the plant managers.

2.2.1 The Qingzhiyuan (Qing) plant

The Qingzhiyuan (Qing) plant serves a residential area in the Xuanwu district of Beijing. The residential area has an area of 300 million m^2. It is stated in *The Regulation of Building Decentralized Wastewater Reclamation Systems in Beijing* that a construction site occupying an area larger than 30,000 m^2 has to build wastewater reuse systems. Accordingly the Qing plant was constructed to accommodate the Qing residential area in 2002. Table 2.1 shows the time progress for the construction of the Qing plant, which illustrates the four years required to complete construction including project application, construction, and equipment installation and operation tests.

Table 2.1
Qing plant construction timeline

Year	Process
2001	Application to build plant
2002	Construction completed
2003	Equipment installation
2004	Operation tests
2005	Operation

The Qing plant was constructed underground and beside the parking place of the residential area of the same name. The depth of the plant is 8 m and its area is 218 m^2. Wastewater is collected separately: grey water is collected through one pipe which is connected with the Qing plant and black water is collected by another pipe which is connected to the municipal sewage system. The grey water covers the shower and sanitation wastewater and the black water covers other wastewater. The decentralized wastewater treatment plant only treats grey water. Figure 2.1 indicates the flow chart of the wastewater reclamation and Table 2.2 illustrates the capacities of the different tanks in the wastewater reclamation plant of the Qing project.

Given the restricted land area in Beijing city centre, all buildings at the Qing plant are high-rise apartment buildings housing 1,100 households, and totaling 11 buildings. Only the buildings constructed after 2000 are equipped with facilities for grey water collection and reclaimed water reuse.

Thus seven out of eleven buildings counting 922 households are equipped with wastewater reuse facilities. The wastewater reuse plant serves only those residences that decide to use reclaimed water and agree to connect their wastewater collection and water supply pipes with the wastewater reclamation plant. According to the interview with the Qing project manager, approximately 80 per cent of households use the reclaimed water. There are 4,910 people living in the residential area, but only 2,500 people use reclaimed water from the Qing plant.

Figure 2.1
Qing plant wastewater reclamation flow chart

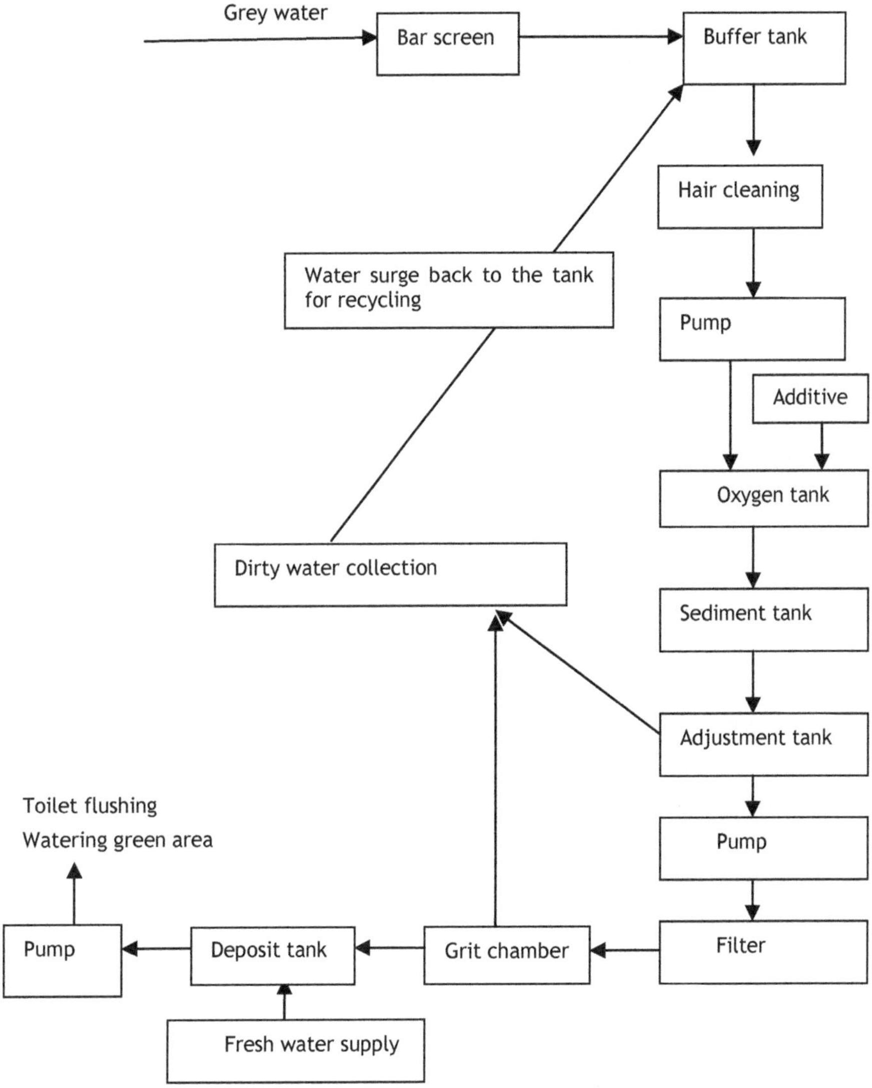

Table 2.2
Capacities of the main facilities at Qing plant

Facilities	Capacity (m^3)
Adjustment tank	200
Deposit tank	80
Oxygen tank	22.5
Sediment tank	10
Buffer tank	7.5

2.2.2 The Beijing Normal University (BNU) plant

The BNU plant serves the Beijing Normal University (BNU) located in the Haidian district of Beijing. The area of the BNU campus is around 700,000 m^2 (source: BNU website). The BNU plant was constructed in 2002 and began operation in 2003.

The BNU wastewater reclamation plant was constructed underground with a depth of around nine metres. Above the plant, there is the office for management, having two floors. The building accounts for an area of 80 square metres. The BNU plant is close to a large campus canteen with a capacity of 500 people. Compared to the noise level at the canteen, the noise generated by the wastewater treatment plant is minimal.

Figure 2.2 indicates the treatment process of the BNU wastewater reclamation plant, which is similar to the treatment technique at the Qing plant. The wastewater is collected from the public shower lounge, and the reclaimed water is reused for toilet flushing in student accommodations and green land irrigation. Because the wastewater is mainly from the public shower lounge, large quantities of hair can be found in the wastewater. A hair-cleaning filter, shown in Figure 2.2, removes the hair that may otherwise block the pipes.

Figure 2.2
BNU plant wastewater reclamation flow char

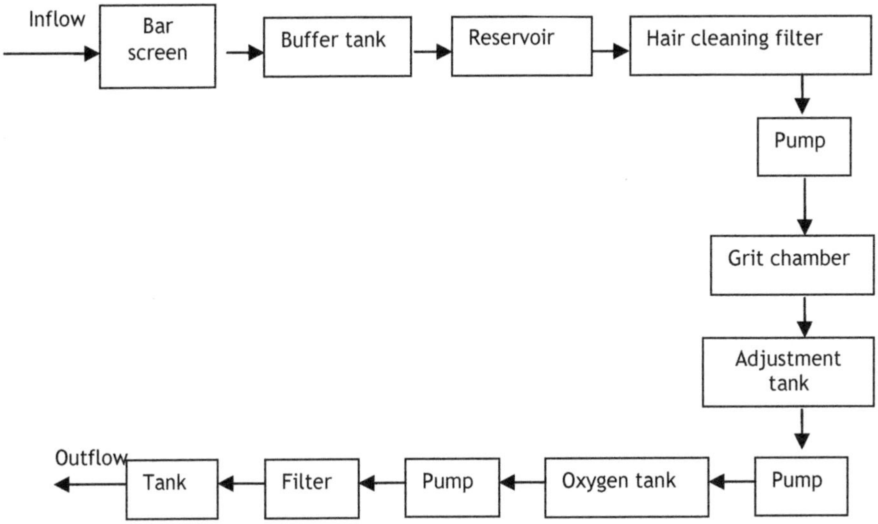

There are almost 30,000 students living on the Beijing Normal University campus. All student accommodations have been equipped with a pipe connection to the BNU plant. The maximum capacity of the plant is 1,000 m^3 per day, but on average, the reclaimed water processed by this plant is around 400 m^3 each day. This is because the quantity of collected wastewater cannot reach 1,000 m^3. New pipes are being built in the student accommodations to collect grey water and will be used in future. As the BNU plant serves students, there is no charge for the reclaimed water. The BNU plant serves all students living on campus.

2.3 Financial Analysis

The financial cost includes initial investment (defined as V_I), operation and maintenance (O&M) cost (defined as $V_{O\&M}$). All components contributing V_I and $V_{O\&M}$ are shown in Equations 2.1 and 2.2, respectively

$$V_I = V_B + V_M + V_P \tag{2.1}$$

$$V_{O\&M} = \sum_{t=1}^{n} \frac{V_t}{(1+r)^t} \tag{2.2}$$

where V_B, V_M and V_P are the initial costs of building construction, electrical and mechanical equipment and pipes, respectively. V_t is the O&M cost occurring in year t; r is the discounted rate; n is the evaluation period (number of years).

As mentioned in Chapter 1, in terms of the *Chinese Economic Evaluation Parameters on Construction* (2006), the discount rate (r) used for the study is eight per cent including the inflation rate. Because few decentralized wastewater reuse systems are operational over a long period in Beijing, the evaluation period (n) is assumed to be ten years.

The financial benefits of a plant are represented by the plant's income, including revenue from reclaimed water charges and subsidies. The Qing plant manager could obtain revenue from reclaimed water charges since the residents pay a fee for reclaimed water. But the BNU plant manager does not have any revenue from reclaimed water charges. The reason is that the BNU plant serves the students of the university who do not need to pay for consumption of reclaimed water. Subsidy is an important source of income for wastewater reuse plants. Generally the Beijing municipal government subsidizes the buildings and equipment of decentralized wastewater reuse plants. In the case of BNU plant, the O&M cost is also subsidized each year. At the Qing plant, only the building of the treatment plant and the equipment was subsidized by the government, while at the BNU plant, the building, equipment and O&M cost are subsidized by the government.

The ratio of financial benefits to financial cost is the criterion to determine the financial feasibility of the plant. If the ratio is larger than one, the plant is financially feasible. Otherwise, the plant is not financially feasible. The financial cost, financial benefits and ratio are calculated in Equations 2.3-2.5, respectively

$$FC_{PV} = V_I + V_{O\&M} \tag{2.3}$$

$$FB_{PV} = \sum_{t=1}^{n} \frac{FB_{r(t)}}{(1+r)^t} + \sum_{t=1}^{n} \frac{FB_{s1(t)}}{(1+r)^t} + FB_{s2} \tag{2.4}$$

$$R_{FB/FC} = \frac{FB_{PV}}{FC_{PV}} \tag{2.5}$$

where FC_{PV} is the financial cost; FB_{PV} is the financial benefits; $FB_{r(t)}$ is the revenue occurring in year t; $FB_{s1(t)}$ is the subsidies occurring in year t; FB_{s2} is the subsidies for initial investment, $R_{FB/FC}$ is the ratio of financial benefits to financial cost.

2.4 Economic Analysis

Table 2.3
Economic, social and environmental effects of decentralized wastewater reuse systems

Economic cost	Initial investment Operation and maintenance cost
Environmental cost	Noise pollution Air pollution
Social cost	Health risk
Economic benefits	Cost savings on constructing pipes Cost savings on water distribution Cost savings on water purification Reuse pollutants
Environmental benefits	Increase in water availability Increase in the water level of rivers Avoidance of overexploitation of water-bearing resources
Social benefits	Raising social awareness

All the economic, social and environmental effects caused by decentralized wastewater reuse systems are listed in Table 2.3, adapted from literature (Hernandez et al. 2006). However, not all the effects listed in Table

2.3 will be included in the economic analysis. Only the major economic, environmental and social effects are selected and quantified using monetary values. The reasons for the selection of only certain effects and the determination of their monetary values are explained below.

First, from the point of view of society, construction, operation and maintenance are seen as consumption of scarce resources, so initial investment and O&M cost are included in the economic cost evaluation, which are the same components contributing to the financial cost.

As there are no traded items in the economic cost and there are no large distortions in market prices of wastewater treatment construction in Beijing, market prices are used for the calculation in this case. Hence the economic cost (defined as V_E) can be obtained by adding the market prices of initial investment (V_I) and O&M cost ($V_{O\&M}$), shown in Equation 2.6.

$$V_E = V_I + V_{O\&M} \tag{2.6}$$

Second, noise and bad smell can be generated during the wastewater treatment processes. The stench can be eliminated through a ventilation system reducing the impact for the inhabitants, while the noise pollution cannot be neglected, as noise is difficult to remove. As the stench does not cause significant effect in this case, air pollution is excluded in the calculation. Only noise pollution is selected as the factor for environmental cost analysis.

Valuation of the effects of noise is very complicated. To simplify the determination, we employ the value used in the literature. Liu (1999) finds that the noise pollution cost in Dalian city is around 108 Yuan per person each year. We calculate the noise pollution cost in the current study by converting the noise pollution value of Dalian City (Liu 1999). The conversion can be made using the differences in income and consumption between Dalian and Beijing city. According to the *Beijing Statistical Yearbook 2005*, the income of Beijing's residents is 1.5 times higher than the income of Dalian's residents. Additionally the ratio of consumption in Beijing to Dalian is also 1.5. It could be assumed that the noise pollution cost of Beijing is 1.5 times higher than the Dalian city. Thus the noise pollution cost per person per year (defined as C_U) in the current study is 162 Yuan. The environmental cost (defined as C_N) can be obtained by multiplying C_U and the number of affected people (defined as N), and is mathematically expressed as:

$$C_N = C_U \times N \tag{2.7}$$

Third, the quantity of pathogens in reclaimed water treated by these small, decentralized plants probably does not reach the official minimum health standards. Human health risks depend on the source of the pathogens, the treatment applied and the exposure route (Ottoson and Stenström 2003). The wastewater reuse plants in this study provide non-potable water for toilet flushing and green land irrigation. The 'spraying irrigation method', which is used by most of the decentralized systems in Beijing, is a typical surface irrigation method. This surface irrigation technique could be negative to human health (Christova-Boal et al. 1996). Thus decentralized wastewater reuse systems in Beijing can have negative effects on human health.

Economists use different methods to value health effects, such as contingent valuation methodology and adjusted human capital methodology. Because of inherent limitations, these economic methods have to be applied to large sample data. We use an indirect valuation method to assess the health effects of wastewater reuse. The Disability Adjusted Life Year (DALY) index is taken as a measurement unit for the effect on human health. DALY is an index of health risk, developed by the World Health Organization (WHO) and the World Bank. One DALY corresponds to one lost year of healthy life and the burden of diseases. It reflects the gap between current health status and an ideal situation where everyone lives with no diseases and disabilities (WHO 2007). DALY is used in many studies for measuring health effects. For example, Aramaki et al. (2006) find that after building wastewater treatment units, the disease burden of a community changed from 60 DALYs per year to 5.7 DALYs per year (Aramaki et al. 2006). In our study, DALY is a bridge to convert the monetary value of health effects from the national level to the scope of a small project. Moreover, in this study, diarrhoea is assumed to be the negative health effect caused by wastewater reuse. Diarrhoea is the largest contributor to the burden of water-related disease (OECD 2007).

The social cost (defined as C_S) is calculated in Equation 2.8. The origin of such calculation method for the social cost is explained as follows. Through the contingent valuation methodology, the World Bank values the total health cost (defined as C_M) caused by water pollution in China, which is about 14.22 billion Yuan each year (World Bank 2007). In terms of the WHO report (2004) figure, the total estimated DALYs (defined as *M*) caused by diarrhoea is 5,055,000 DALYs each year. The DALY cost rate (C_M / M) namely the cost per DALY, is calculated at 2,813 Yuan per DALY per year. The product of DALYs rate (defined as *R*) and population number (defined as *K*) gives the total DALYs for Beijing. As a result of missing data, the Beijing DALY rate (*R*) is determined by the DALY rate of China, which is 389×10^{-5} DALYs per person (WHO 2004). The registered permanent population living in Beijing's central district is 2.25 million. It is supposed that the DALYs of Beijing resulting in diarrhoea is caused entirely by reclaimed wastewater. Accordingly the probability of DALYs due to irrigating reclaimed water on green land (P_1) could be represented by the ratio of reclaimed water amount for green area irrigation to the total reclaimed water amount in Beijing. P_2 denotes the probability of DALYs due to irrigating the project's green land. Since large area of green land surface could increase the infection of diarrhoea, P_2 is represented by the ratio of green land area in the project to total green land surface of the Beijing city centre.

$$C_S = \frac{C_M}{M} \times R \times K \times P_1 \times P_2 \tag{2.8}$$

Fourth, as listed in Table 2.3, the economic benefits generally include cost savings on constructing pipes, water purification and distribution, and reuse of pollutants. Being the conventional system, centralized wastewater reuse systems have been in place in Beijing for many years, and require substantial investments in pipe construction for reclaimed water distribution, due to the long distance between centralized plants and users. Compared with centralized wastewater reuse systems, decentralized systems require shorter reclaimed water distribution pipes thereby saving on pipe construction. As the capacity of a decentralized plant is usually limited, the cost savings on water purification and distribution is so small that it can be ignored in the current analysis. Generally the pollutants of decentralized wastewater reclamation are not reused in the Beijing urban areas, so the benefit of reuse of pollutants is not considered in the study. As a result, only the cost savings on pipe construction is selected for the economic benefits analysis. Cost savings on water purification and distribution, and reuse of pollutants are neglected in the economic benefits analysis.

Figure 2.3
Location of Beijing centralized wastewater reclamation plants and the two decentralized plants studied

There are five large centralized plants in Beijing: Gaobeidian, Fangzhuang, Wujiacun, Qinghe and Jiuxianqiao. The Fangzhuang wastewater reclamation plant shown in Figure 2.3 is the closest to the Qing plant, and the Jiuxianqiao plant is the closest to the BNU plant. We assume that the closest centralized plant would provide the reclaimed water if there is no on-site plant. Hence the economic benefits of avoiding constructing pipes (defined as B_L) can be calculated as:

$$B_L = C_L \times L \quad (2.9)$$

where C_L is construction cost per metre pipe and L is the distance between the closest centralized plant and the studied plants.

According to interviews with officials of the Beijing drainage group, the value of C_L is between 2,000 and 20,000 Yuan/m. We take the least unit cost value 2,000 Yuan/m and the shortest distance between the on-site plant and the closest big plant for the estimation.

Fifth, increasingly 'new water' is created through reusing wastewater, decreasing the stress on water resources depletion. The increase in water availability is a crucial environmental benefit, especially for a city like Beijing, which has water scarcity. However, based on the two plants studied, the actual increase in the river water level and reduction of the overexploitation of water-bearing resources cannot be recognized. For simplicity the current study assumes that only the increase in water availability makes major contributions to the environmental benefits.

The shadow price of Beijing water resources is estimated to be around three Yuan/m^3 (Liu and Chen 2003). The environmental benefit (defined as BE) of increase in water availability can be calculated by Equation 2.10.

$$B_E = C_E \times E \tag{2.10}$$

Where C_E is unit water monetary value and E is the amount of reclaimed water.

Finally, increasing public awareness on utilizing reclaimed water is still a long way off. Normally awareness improvement could be reached through various public education and advertisement campaigns. The introduction of decentralized wastewater reuse systems is a method to enhance awareness concerning water saving so that cost is saved on awareness rising campaigns. It is assumed that the educational effect of a decentralized plant is the same as the effect of a public campaign. The cost savings on campaigns can be regarded as the social benefits (defined as B_S) of the wastewater reuse plants. This can be determined by total expenditure on public awareness raising campaign (defined as S) and the ratio of number of users to total population in Beijing (defined as Q) as expressed in Equation 2.11.

$$B_S = S \times Q \tag{2.11}$$

The average cost of public campaign (S) in water sector in Beijing is 2,780,000 Yuan/year (DPP 2001).

All the parameters used to determine the monetary values of economic, environmental and social effects are summarized in Table 2.4.

The ratio of benefits to cost (defined as $R_{B/C}$) is used as the criterion for economic feasibility. If $R_{B/C} > 1$, the plant is economically feasible. If $R_{B/C} < 1$, it means the plant is not economically feasible. The cost (C_{PV}), benefits (B_{PV}) and the ratio of benefits to cost ($R_{B/C}$) are calculated by Equations 2.12-2.14, respectively.

$$C_{PV} = V_E + \sum_{t=1}^{n} \frac{C_N}{(1+r)^t} + \sum_{t=1}^{n} \frac{C_S}{(1+r)^t} \tag{2.12}$$

$$B_{PV} = B_L + \sum_{t=1}^{n} \frac{B_E}{(1+r)^t} + \sum_{t=1}^{n} \frac{B_S}{(1+r)^t} \tag{2.13}$$

$$R_{B/C} = \frac{B_{PV}}{C_{PV}} \tag{2.14}$$

It is assumed that the values of environmental cost (C_N), social cost (C_S), environmental benefit (B_E) and social benefit (B_S) in each year do not change during the evaluation period.

Table 2.4
Summary of parameters on determination of cost and benefits of decentralized wastewater reuse systems

Parameter	Definition
V_I	Initial investment (Yuan)
$V_{O\&M}$	Operation and maintenance cost (Yuan)
V_E	Economic cost (Yuan)
C_U	Unit cost of noise effect (Yuan per person per year)
N	Affected user number (persons)
C_N	Environmental cost (Yuan/year)
C_M	Total health cost (billion Yuan/year)
M	Total DALYs caused by water (DALYs/year)
R	DALYs rate (DALYs per person per year)
K	Population of Beijing (million persons)
P_1	Probability of DALYs due to irrigating reclaimed water on green land (%)
P_2	Probability of DALYs due to irrigating the green land of the plant (%)
C_S	Social cost (Yuan/year)
C_L	Unit cost on pipes construction (Yuan/m)
L	Distance between closest centralized plant and users (m)
B_L	Economic benefit (Yuan)
C_E	Water monetary value (Yuan/m^3)
E	Amount of reclaimed water (m^3/year)
B_E	Environmental benefit (Yuan/year)
S	Total spent on public awareness raising campaign (Yuan/year)
Q	Ratio of number of users to total population (%)
B_S	Social benefit (Yuan/year)

2.5 Results

2.5.1 Results of economic and financial analysis

Table 2.5 presents the results of the financial analysis of both plants. It shows that total initial investments are 2.9 million Yuan in the Qing plant and 3.7 million Yuan in the BNU plant. Although the treatment capacity of the BNU plant is almost seven times larger than that of the Qing plant, the difference in the initial investment values between two plants is not significant.

In the O&M cost, electricity cost is much higher than the other O&M costs. For example, the electricity consumption of the BNU plant each year is around 131,765 Yuan. The personnel cost being the second largest cost in O&M, is only one-third of the electricity cost. The electricity cost depends on the capacity of the plant and the unit cost of energy. Hence the capacity of the plant and the unit cost of energy have significant influence on the O&M cost of a wastewater reuse plant.

Table 2.5
Financial analysis of decentralized wastewater reuse systems

	Qing plant	BNU plant
Financial cost		
Initial investment (Yuan)		
Buildings	40,000	100,000
Equipment	260,000	500,000
Pipes	2,600,000	3,100,000
Sub-total	2,900,000	3,700,000
O&M cost (Yuan/year)		
Electricity	45,638	131,765
Chemical	7,000	10,000
Maintenance	1,200	12,235
Personnel	27,000	46,000
Sub-total	80,000	200,000
Financial Benefits		
Revenue (Yuan/year)	21,000	0
Subsidies (Yuan)	300,000	1,942,000

For the sake of comparative analysis, the present values of all effects in the economic analysis are calculated and listed in Table 2.6. The environmental cost of the Qing plant is 32,611 Yuan whereas the environmental benefits of the Qing plant are 402,605 Yuan. This means that the environmental benefits of the Qing project are 12 times larger than the environmental cost. For the BNU plant, the environmental benefits are 260 times larger than the environmental cost. Hence the positive environmental effects caused by the decentralized wastewater reuse plants are larger than the negative environmental effects.

Table 2.6
Economic analysis of decentralized wastewater reuse systems

	Qing plant	BNU plant
COST		
Economic cost (Yuan)	3,437,000	5,042,000
Environmental cost (Yuan)	32,611	10,870
Social cost (Yuan)	13,212	13,212
TOTAL	3,482,823	5,066,082
BENEFITS		
Economic benefits (Yuan)	16,000,000	24,000,000
Environmental benefits (Yuan)	402,605	2,818,000
Social benefits (Yuan)	21,411	290,000
TOTAL	16,424,016	27,108,000

The economic benefits are represented by the value of cost savings on constructing pipes, accounting for around 90 per cent of total benefits. In centralized systems, the reclaimed water distribution pipes would have to be built in existing urban areas through demolition and relocation, leading to extremely high costs in pipe construction. The cost of pipe construction could be effectively saved by decentralized systems. In the Qing plant, cost savings on constructing pipes is 16 million Yuan, whereas, initial investment in the Qing plant was only 2.9 million Yuan. In the BNU plant, cost savings on pipes is 24 million Yuan and initial investment at the BNU plant was 3.7 million Yuan. This implies that the funding for pipe construction to distribute reclaimed water could finance the investments of five or six decentralized plants.

Table 2.7 shows the results of the financial and economic feasibility analysis. In the economic analysis, the ratio of benefits to cost of the Qing plant is 4.7, which is greater than one. The ratio of the BNU plant is also greater than one. This shows that both Qing and BNU plants are economically feasible, which indicates that decentralized wastewater reuse systems have positive effects on the welfare of the society. From the point of view of the government, decentralized wastewater reuse systems deserve to be promoted. However, in the financial analysis, the ratios of financial benefits to cost of both plants are less than one, which implies that the two plants are not financially feasible. Thus the plant managers would prefer not to operate the wastewater reuse systems and the systems may not remain operational in the long term.

Table 2.7
Results of financial and economic feasibility of decentralized wastewater reuse systems

	Qing plant	**BNU plant**
Financial analysis (ratio of financial benefits to financial cost: $R_{FB/FC}$)	0.13	0.38
Economic analysis (ratio of benefits to cost: $R_{B/C}$)	4.7	5.4

2.5.2 Further discussion of the results

For the sake of systematic analysis, the researcher created a coding form shown in Table 2.8 (Lipsey and Wilson 2001). The information and evaluation results are coded as either 0 or 1. Table 2.8 shows that the Qing plant has a different score as the BNU plant only on item A.

The scores on item A imply that the BNU plant has a much larger capacity than the Qing plant. It was found that there was economic scale in wastewater reclamation and reuse, namely the unit cost decrease when the system scale becomes larger (Friedler and Hadari 2006; Yamagata et al. 2003). Economies of scale imply that decentralized treatment systems may have a higher unit cost than the centralized system does. The unit O&M cost of the Qing plant is higher than that of the BNU plant. However, no matter the scale, both plants studied show the same results: economically feasible but not financially feasible. Hence the economic feasibility or financial feasibility is not related to the scale of operation according to this study.

Table 2.8
Codified data for the Qing and BNU plants

	Qing plant	BNU plant
A. Economic scale	0	1
B. Unit O&M cost	1	1
C. Total cost recovery	0	0
D. Financial feasibility	0	0
E. Economic feasibility	1	1

A, 0: small; 1: large

B, 0: unit O&M cost is smaller than reclaimed water rate; 1: unit O&M cost is larger than reclaimed water rate

C, 0: total cost is not recovered; 1: total cost is recovered

D, 0: not financially feasible; 1: financially feasible

E, 0: not economically feasible; 1: economically feasible

The scores on item B indicate that the unit O&M costs of two plants are higher than the rate for reclaimed water. The unit O&M cost of the Qing plant is around 3.8 Yuan/m^3 and the unit O&M cost of the BNU plant is around 1.5 Yuan/m^3. The reclaimed water rate is one Yuan/m^3, which is much lower than the O&M cost. The rate for reclaimed water determines the financial benefits of a plant and the low rate affects the cost recovery in a negative way. Item C shows that total cost of both plants cannot be recovered financially. The low rate of reclaimed water is an important factor that does not contribute to cost recovery, thereby leading to the decentralized wastewater reuse system not being financially feasible.

As the quality required for reclaimed water is lower than the quality required for drinking water, there is a misconception that the cost of reclaimed water is lower than that of drinking water. Although the cost of tertiary treatment for reclaimed water is low, the cost of reclaiming wastewater in an entire treatment process is high (Angelakis et al. 2003; Borboudaki et al. 2005; Ogoshi et al. 2001). For example, the study of Ogoshi et al. (2001) indicates that the cost of reclaimed water in the Fukuoka City of Japan is 2.01 US dollar/m^3, while the cost of drinking water is only 1.88 US dollar/m^3. Following those findings in literature, it is assumed that the cost of reclaimed water in Beijing is also higher than the drinking water. The price of reclaimed water is fixed at one Yuan/m^3 by the government whereas the price of drinking water is 3.7 Yuan/m^3. This implies that the current rate of one Yuan/m^3 on reclaimed water does not reflect the real cost.

This shows that economic scale is not implicated in lack of financial feasibility. The low rate charged for reclaimed water is the crucial factor as to why decentralized water reuse plants are not financially feasible. The reclaimed water rate is lower than the actual O&M cost and does not reflect the real cost of reclaimed water.

2.6 Conclusions

The chapter evaluates the decentralized wastewater reuse systems in Beijing through an integrated financial and economic feasibility analysis. The financial analysis is made from the point of view of project manager, while the economic analysis is from the point of view of society. The major economic, environmental and social effects of the plants are all considered in the economic analysis.

The analysis indicates that decentralized wastewater reuse systems are economically feasible. It means that the systems have positive effects on society. Thus, from the point of view of government or society, the decentralized wastewater reuse systems are worth promotion.

However, the decentralized wastewater reuse systems are not financially feasible. This implies that there are serious financial problems in the systems. The low rate charged for reclaimed water is the key reason for the systems not being financially feasible. From the plant manager's perspective, the decentralized systems may not continue to operate in the long term if the financial problems are not solved. Thus solving the financial problems of decentralized wastewater reuse systems should be on the political agenda in the future (Angelakis et al. 2003). It would require subsidies unless realistic pricing policies for water are introduced.

Notes

[1] Chapter 2 is published in the journal, *Water Science and Technology* 61(8): 1965-73 (2010).

3 Financial and Economic Analysis of Centralized Wastewater Reuse Systems and Comparison Between Centralized and Decentralized Systems[2]

3.1 Introduction

Centralized wastewater treatment system, as the conventional system, has been applied over many decades in developed countries. Normally the term centralized wastewater treatment system is used to describe systems consisting of a sewer system that collects wastewater from households, small enterprises, industrial plants and governmental buildings, and transports this ever-changing mixture to a wastewater treatment plant. But this concept is increasingly challenged, especially in developing countries.

The deficiencies of centralized approaches highlight the potential benefit of adopting a decentralized approach to urban water management recently. Because the scale of decentralized systems is relatively small, the initial investment of a decentralized plant is comparatively low and the management system is relatively simple. Generally decentralized systems are less resource intensive and a more ecologically benign form of wastewater treatment and sanitation (Lens et al. 2001). In contrast to centralized systems, decentralized systems render service close to the origin. In the decentralized systems, the wastewater is collected and transferred to a plant, and then the treated water is reused and the remaining sludge is converted into fertilizer.

Many researchers study whether decentralized water treatment systems are a good alternative to centralized systems. Some carry out the comparison between centralized and decentralized water treatment systems from the point of view of technology (Norton 2009; Wilderer and Schreff 2000), and others make the comparison from the perspective of cost (Fane et al. 2002; Gratziou et al. 2005; Jia et al. 2005; Wang et al. 2008). Rarely centralized and decentralized wastewater reuse systems are compared from the perspectives of economics.

This study conducts a comparative analysis between the centralized and decentralized wastewater reuse systems in Beijing. The comparison focuses on the financial and economic feasibility and the environmental and social effects. The objective of this chapter is to discover whether the centralized wastewater reuse systems are economically and financially competitive with the decentralized systems.

Chapter 2 examined the decentralized wastewater reuse systems in Beijing and concluded that the decentralized wastewater reuse systems are economically feasible but not financially feasible. Subsequently this chapter studies the centralized wastewater reuse systems in Beijing. Based on the analytical framework shown in Chapter 1 (Figure 1.17), an analysis of the financial and economic feasibility of the centralized wastewater reuse systems in Beijing is carried out. The financial analysis and economic analysis are performed separately from the different perspectives of various decision makers. The financial analysis is from the perspective of plant manager while the economic analysis is from the perspective of society. After the financial and economic feasibility analysis, a comparative study was carried out between centralized and decentralized wastewater reuse systems.

In this chapter, Section 3.2 introduces the centralized wastewater reuse plants in Beijing. Section 3.3 presents the financial and economic analysis of the centralized wastewater reuse plants. Section 3.4 shows the comparative analysis between centralized and decentralized systems. Finally, the conclusion of this study is in section 3.5.

3.2 About the Centralized Wastewater Reuse Plants

The Gaobeidian wastewater reuse plant is the first centralized wastewater reuse plant in Beijing, constructed in 2000, while the first decentralized wastewater reuse plant was built in 1987. Compared with the decentralized wastewater reuse systems, the centralized systems are relatively new in Beijing. The total expected amount of the reclaimed water of the centralized wastewater reclamation plants is 0.66 million m^3 per day. However the reclaimed water is less than 60 per cent of the expected amount.

Figure 3.1
Location of Beijing centralized wastewater reclamation plants

As indicated in Figure 3.1, there are five centralized wastewater reclamation plants in operation or still under construction in Beijing. Only the Gaobeidian plant and the Jiuxianqiao plant have operated for several years. Hence the Gaobeidian plant and the Jiuxianqiao plant were chosen for the economic and financial analysis.

3.2.1 The Gaobeidian (Gao) plant

The Gaobeidian (Gao) plant is the largest wastewater reuse plant and the first centralized wastewater reclamation plant in Beijing, operated since 2000. The design capacity of the Gao plant is 470,000 m^3/day, at present the volume of the reclaimed water processed by the Gao plant is 300,000 m^3/day. The length of pipes for water distribution in the Gao plant is around 24 kilometres (Beijing Water Authority 2004). The Beijing drainage group is in charge of the Gao project. The main stakeholders of the Gao plant appear in Chapter 1, they include the Drainage Group, the Municipal Environmental Protection Bureau, the Municipal Administration Committee and the Urban Construction Bureau (Figure 1.14).

Figure 3.2
Distribution of reclaimed water from Gao plant

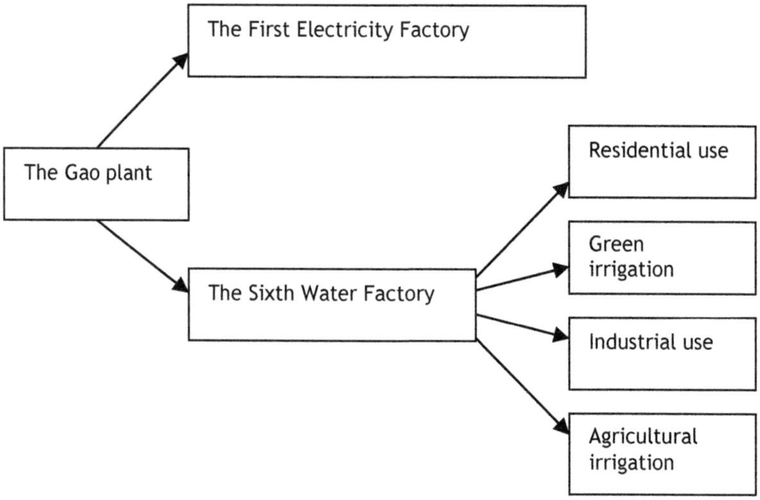

The planned use for reclaimed water from the Gao plant is for agricultural irrigation, industrial cooling water, green lands irrigation, residential toilet flushing and car washing (shown in Figure 3.2). Table 3.1 indicates quantity distribution of reclaimed water from the Gao plant. Around 200,000 m^3 of reclaimed water from the Gao plant goes to 'The First Electricity Factory' for cooling water each day, which accounts for around 68 per cent of total quantity of reclaimed water. About 100,000 m^3 of water is delivered to 'The Sixth Water Factory' each day to be processed again for water quality improvement and then the water goes to residential use, green irrigation, industrial use and agricultural irrigation (Table 3.1). The residential use (shown in Table 3.1) means using reclaimed water for toilet flushing and green irrigation in residential areas.

Table 3.1
Quantity distribution of reclaimed water from the Gao plant

Use	Quantity (m^3/day)
Cooling water for electricity factory	200,000
Residential use	20,000
Green irrigation	30,000
Industrial use	20,000
Agricultural irrigation	30,000
Total	300,000

3.2.2 The Jiuxianqiao (Jiu) plant

The Jiuxianqiao (Jiu) plant was completed in 2004. The amount of reclaimed water processed by the Jiu plant is 60,000 m^3 per day, of which 19,200 m^3 is for residential use and 30,000 m^3 is for lake or river water supplementation, 7,800 m^3 is for green irrigation and the remaining 3,000 is for agricultural irrigation (shown in Table 3.2). This illustrates that the reclaimed water from the Jiu Plant is mostly used for residence and water supplementation. Compared with other uses, the amount of reclaimed water for agricultural irrigation is much smaller. The length of the pipes for water distribution in the Jiu plant is around 17 kilometres (Beijing Water Authority 2004). Like the Gao plant, the Jiu plant is also owned by the Beijing Drainage Group. So the main stakeholders of the Jiu plant include the Drainage Group, the Municipal Environmental Protection Bureau, the Municipal Administration Committee and the Urban Construction Bureau (Figure 1.14, Chapter 1).

Table 3.2
Quantity distribution of reclaimed water from the Jiu Plant

Use	Quantity (m^3/day)
Residential use	19,200
Lake or river water supplementation	30,000
Green irrigation	7,800
Agricultural irrigation	3,000
Total	60,000

3.3 Financial and Economic Analysis of Centralized Wastewater Reuse Systems

Based on the evaluation framework of Chapter 1, the analysis of financial and economic feasibility of the centralized wastewater reuse systems appears in this section. The financial analysis is carried out from the point of view of the plant manager, while the economic analysis is from the viewpoint of the government. The detailed explanation and the results of the financial and economic analysis are now presented.

3.3.1 Financial analysis

The financial analysis contains evaluations of financial costs and financial benefits. The financial costs include initial investment (defined as U_I), operation and maintenance (O&M) cost (defined as $U_{O\&M}$). All components contributing U_I and $U_{O\&M}$ are shown in Equations 3.1-3.2, respectively

$$U_I = U_C + U_D + U_P + U_R + U_O \tag{3.1}$$

$$U_{O\&M} = \sum_{t=1}^{m} \frac{U_t}{(1+r)^t} \tag{3.2}$$

where U_C, U_D, U_P, U_R, and U_O are the initial costs of construction, demolition and relocation, preparation, interest and others, respectively. U_t is the O&M cost occurring in year *t*; *r* is the discounting rate; *n* is the evaluation period (number of years).

According to the publication *Chinese Economic Evaluation Parameters on Construction* (2006), the discount rate (*r*) used for cost benefits studies in China is eight per cent including the inflation rate. Inflation rates in China for years 2007 and 2008 were 4.8 per cent and 5.9 per cent, respectively, hence the opportunity cost of capital is around three per cent (China Statistical Yearbook 2007, 2008). The evaluation period (*n*) is assumed to be 20 years.

The financial benefits of a plant are represented by the income from the plant, including revenue from reclaimed water charges and subsidies. In terms of its usage, the reclaimed water is charged at different rates. Table 3.3 shows the latest reclaimed water rates in Beijing. Based on the data from tables 3.1, 3.2 and 3.3, the revenue of the Gao plant and the Jiu plant can be calculated. Subsidy is another important source of income for wastewater reuse plants. In this study, only the Gao plant received subsidies on the initial investments from the Beijing municipal government.

Table 3.3

Charge rates for reclaimed water (Yuan/m^3)

Usage	Domestic	Landscape	Industrial	Power generating plant
Charge rates	1.2	0.8	1.5	0.9

The ratio of financial benefits to financial cost is the criterion to determine the financial feasibility of the plant. If the ratio is larger than 1, the plant is financially feasible. Otherwise, the plant is not financially feasible. The financial cost, financial benefits and ratio are calculated by Equations 3.3-3.5, respectively

$$fc_{pv} = U_I + U_{O\&M} \tag{3.3}$$

$$fb_{pv} = \sum_{t=1}^{n} \frac{fb_{r(t)}}{(1+r)^t} + fb_s \tag{3.4}$$

$$r_{fb/fc} = \frac{fb_{pv}}{fc_{pv}} \tag{3.5}$$

where fc_{pv} is the financial cost; fb_{pv} is the financial benefits; $fb_{r(t)}$ is the revenue occurring in year t; fb_s is the subsidies for initial investment, $r_{fb/fc}$ is the ratio of financial benefits to financial cost.

3.3.2 Economic analysis

Table 3.4
Economic, social and environmental effects of centralized wastewater reuse systems

Economic cost	Initial investment Operation and maintenance cost
Environmental cost	Carbon dioxide emission Noise pollution Air pollution
Social cost	Health risk Residential resettlement
Economic benefits	Cost saving on fertilizers Reuse of pollutants
Environmental benefits	Increase of water availability Increase in the level of rivers Avoidance of overexploitation of water-bearing resources
Social benefits	Raising social awareness Improving employment

As shown in the evaluation framework (Figure 1.17, Chapter 1), the economic, environmental and social effects are all taken into consideration in the economic analysis. In terms of literature and interviews with officials at Beijing Water Authority, all the economic, social and environmental effects caused by centralized wastewater reuse systems are listed in Table 3.4 (Asano 2005; Beijing Water Authority 2002; Hernandez et al. 2006). Similar to the economic analysis of Chapter 2, not all the effects listed in Table 3.4 will be included in the economic analysis. Only the major economic, environmental and social effects are selected and quantified using monetary values. The reasons for selecting certain effects and the determination of their monetary values are explained below.

First, from the point of view of society, construction, operation and maintenance are seen as consumption of scarce resources, so initial investment and O&M cost are included in the economic cost evaluation, which are the same components contributing to the financial cost.

Because few items are traded in the initial investment and the O&M cost for urban wastewater treatment, and because there are no large distortions in market prices of wastewater treatment construction in Beijing, market prices could be used directly for the calculation of economic cost in the study. The economic cost (defined as c_e) can be obtained by adding the present values of initial investment (u_n) and O&M cost ($u_{O\&M}$), shown in Equation 3.6.

$$c_e = u_n + u_{O\&M} \tag{3.6}$$

Second, as centralized wastewater reuse systems generally have large-scale wastewater treatment plants, the energy consumption for wastewater reclamation is vast. There is high carbon dioxide emission during the process of energy generation, and carbon dioxide emission is a greenhouse gas causing a negative environmental impact. Hence large energy consumption could result in negative environmental influences. Moreover, noise and malodors generated during the wastewater treatment processes may lead to environmental cost. However, because centralized systems are generally constructed in suburban areas with fewer residents, the noise and stench generated by wastewater reclamation cause fewer effects so the effects of noise and smell are not considered in this study. Therefore only the environmental cost of carbon dioxide emission is considered in the analysis of environmental cost.

Although there are other emissions during the process of energy generation such as sulphur dioxide and nitrogen oxides, which also cause negative environmental effects, the amount of carbon dioxide emission accounts for around 60 per cent (Skeer and Wang 2005). To simplify the study, only the effect of carbon dioxide was considered. The energy generation could be coal power, petroleum power, gas power and wind power. In China, 70 per cent of energy consumption is from coal power and the majority of carbon dioxide emissions result from coal combustion (Dahowski et al. 2009). So we assume that 70 per cent of the energy for wastewater reclamation in centralized wastewater reuse systems in Beijing is from coal power. Due to smaller carbon dioxide emissions, the effect caused by the remaining 30 per cent of energy generation is neglected. Hence only the carbon dioxide from coal power is calculated for the environmental cost.

The primary environmental effect caused by carbon dioxide emission is climate change, which is an international issue at present. But the social impact of climate change is complex and difficult to evaluate because of the uncertainty of climate change. In the literature, the indicators considered in various studies are different and even the values for parameters such as discounting rate used for discounting the life of carbon dioxide are also different (Skeer and Wang 2005; Tol 2005, 2008). The social impact of climate change could be regarded as the damage cost of carbon dioxide. Tol (2005) finds that the estimated results of the damage cost of carbon dioxide in existing studies have a large difference, ranging from $5/ton to $400/ton carbon dioxide. Tol calculates the mean value of marginal damage cost in the peer-reviewed literature, which is $50/ton carbon dioxide (2005). So we take the mean value of marginal damage cost $50/ton (350 Yuan/ton) carbon dioxide as unit environmental cost of carbon dioxide (defined as u_d) in the study. According to the literature, the unit carbon dioxide emission (defined as d) in a coal power plant is around 800g/kWh (Skeer and Wang 2005). So the total amount of carbon dioxide emission due to energy consumption in centralized wastewater reuse systems could be obtained by multiplying 70 per cent of the energy consumption (defined as g) and carbon dioxide emission (d). Hence the environmental cost (defined as c_d) can be obtained by multiplying the unit environmental cost per carbon dioxide emission (u_d) and the amount of carbon dioxide emission, and is mathematically expressed as

$$c_d = u_d \times g \times d \tag{3.7}$$

Third, if the reclaimed water reaches the standard of water quality, it has few negative effects on human health. Normally state owned water companies manage the centralized wastewater reclamation plants in China, so theoretically the reclaimed water quality of centralized plants reaches the standard requirements and the quantity of pathogens in reclaimed water reaches the official minimum health standards. There is a low probability that centralized wastewater reuse systems cause health risk. Hence the health risk is not calculated in the social cost. Pipe construction is an important and difficult part of the construction of a centralized wastewater reuse plant. In addition to a high expenditure on pipe construction, there are serious social influences caused by pipe construction. According to an interview with the officials of Beijing Water Authority, 60 per cent of the distribution pipes of centralized wastewater reuse systems are constructed through demolition and relocation and 40 per cent of pipes are constructed following rivers. There are severe influences on society when pipes are constructed through demolition and relocation, while there are slight influences when pipes are constructed following rivers. So this study only focuses on the social effects of demolition and relocation.

Demolition and relocation can lead to changes in the road net, the destruction of existing city buildings and residential resettlement (Camagni et al. 2002). Because of the extreme difficulty of quantifying social cost, it is rare in the literature that the total social costs caused by demolition and relocation in pipe construction are evaluated. Most research focuses on financial costs such as the compensation to people resettled in other living places (Malpezzi 1999; Mao 1966; Osman et al. 2008). It is worth noting that this study only evaluates the main social costs resulting from the major effects. This could help to simplify the economic determination of social cost of demolition and relocation due to pipe construction.

Among the effects caused by demolition and relocation, the effect of residential resettlement is the most important, because residential resettlement can lead to people becoming unemployed or increased cost in education, medical health care and transportation (Luo 2007). The effect of residential resettlement is the most important effect considered in this study. Residential resettlement means, residents living along the line of pipe construction must move to accommodate construction. As mentioned previously, residential resettlement can result in various negative influences on people's life, such as unemployment and increasing living costs. It seems to be very difficult to evaluate all the negative influences from residential resettlement. In the literature, the increased transportation cost due to residential resettlement is an essential and real effect on people's lives (Camagni et al. 2002; Luo 2007). So, only the increased cost of transportation due to residential resettlement is regarded as social cost of demolition and relocation and determined in this study.

Equation 3.8 shows the determination of the increased cost of transportation due to residential resettlement, namely the social cost of demolition and relocation (defined as c_s).The origin of such calculation is as follows. The increased cost of transportation can be calculated by multiplying the average increased public transport cost for one person (defined as μ_p) and the affected number of people. Since public transportation is the main travel method for Beijing residents, after moving to a new place, it is assumed that all affected residents will make one additional transfer each day via public transportation. The transportation fees due to the travel transfer could be regarded as an increased transport cost. The average public transportation (including metro and bus) cost in Beijing is around 4 Yuan per person for a round trip. The population density (defined as p) and the area calculate the population affected by the resettlement. The area can be obtained by multiplying the length of pipe construction (defined as l) and the width, which is supposed to be one metre. Hence the product of the population density (defined as α) and the length of pipe construction (defined as l) can be regarded as the number of people affected by the resettlement. In the Beijing Statistic Yearbook (2009), the population density of urban areas of Beijing is 20,000 persons per square metre.

$$c_s = \mu_p \times p \times l \tag{3.8}$$

Fourth, part of the reclaimed water produced by centralized wastewater reuse systems is reused for agricultural irrigation. The reclaimed water contains nitrogen and phosphorus, which are important fertilizers for agricultural production (Wang 2007). Using reclaimed water for agricultural irrigation limits the use of fertilizers in agricultural production. The cost saving on fertilizers could be regarded as economic benefits of centralized wastewater reuse. Moreover, since normally the pollutants of wastewater reuse systems in Beijing are not reused, the benefit of reusing pollutants is not considered in the study. The economic benefit of cost saving on fertilizers (defined as b_f) could be determined by multiplying the unit cost of saving on fertilizers (defined as u_f) and the amount of reclaimed water for agricultural irrigation (defined as f), shown in Equation 3.9. According to the results of a working paper of the Beijing Water Authority, the unit cost saving on fertilizers due to using reclaimed water for agricultural irrigation is 0.0225 Yuan per m^3 (Beijing Water Authority 2002).

$$b_f = u_f \times f \tag{3.9}$$

Fifth, table 3.4 shows that the environmental benefits include increase in water availability and river levels, as well as avoidance of overexploitation of water-bearing resources. Since large quantities of 'new water' is created by reclaiming wastewater and is reused, 'increase of water availability' is an important environmental benefit. But there are no significant effects of 'increase in the level of rivers' and 'avoidance of overexploitation of water-bearing resources' caused by centralized wastewater reuse systems. Although around half of the reclaimed water produced by the Jiu plant is used to supplement the lakes and rivers, the quantity is too small to take into consideration. Therefore only the 'increase of water availability' makes a major contribution to the environmental benefits. Equation 3.10 calculates the environmental benefit of increase in water availability

$$b_e = u_e \times e \tag{3.10}$$

where b_e is the environmental benefit, u_e is the monetary value of water, e is the amount of reclaimed water. In the literature, the monetary value of water resources (u_e) is estimated to be around 3 Yuan/m^3 in Beijing (Liu and Chen 2003).

Finally, as presented in Chapter 2, the introduction of using reclaimed water is a method to improve public awareness concerning water saving. However, in centralized systems, the reclaimed water could be obtained easily through direct distribution, which is similar to accessing tap water. Accordingly people may not realize that they are using reclaimed water. Public awareness of using reclaimed water and saving water is not influenced. Hence raising social awareness is not considered in the determination of social benefit in the study.

Rapid economic growth creates employment (Li 2003; Rawski 1979). The operation of a centralized wastewater reuse systems generally needs many workers, which creates new jobs. This implies that the operation of a centralized wastewater reuse system can help improve employment in the region. It is assumed that the employment effect of a centralized wastewater reuse system is the same as the effect of economic growth. The value of economic growth contributing to improving employment can be regarded as the social benefit of improving employment caused by centralized wastewater reuse systems.

Equation 3.11 indicates the determination of social benefit of improving employment. It is defined that w is the number of workers of the plant, W is the total employment of the region, Y is the GDP of the region, b_s is the value of economic growth namely social benefit of employment improvement, and β is the employment elasticity. The employment elasticity (β) means the ratio of the increase of employment (namely $\frac{w}{W}$) to the increase of economic growth (namely $\frac{b_s}{Y}$). In terms of literature, the employment elasticity of China is estimated to be 0.3, which means that if the economic growth increases 1 per cent, there will be an increase of 0.3 per cent in employment (Li 2003).

$$b_s = \frac{\frac{w}{W}}{\beta} \times Y \qquad (3.11)$$

All the parameters used to determine the monetary value of economic, environmental and social effects are summarized in Table 3.5.

The ratio of benefits to cost (defined as $r_{b/c}$) is used as the criterion for economic feasibility. If $r_{b/c} > 1$, the plant is economically feasible. If $r_{b/c} <$ 1, it means the plant is not economically feasible. Equations 3.12, 3.13 and 3.14 respectively, calculate the present value of the cost (c_{pv}) and benefits (b_{pv}) and the ratio of benefits to cost ($r_{b/c}$).

$$c_{pv} = c_e + \sum_{t=1}^{n} \frac{c_d}{(1+r)^t} + \sum_{t=1}^{n} \frac{c_s}{(1+r)^t} \qquad (3.12)$$

$$b_{pv} = \sum_{t=1}^{n} \frac{b_f}{(1+r)^t} + \sum_{t=1}^{n} \frac{b_e}{(1+r)^t} + b_s \tag{3.13}$$

$$r_{b/c} = \frac{b_{pv}}{c_{pv}} \tag{3.14}$$

Table 3.5
Summary of the parameters on determination of cost and benefits of centralized wastewater reuse systems

Parameter	Definition
u_n	Initial investment (Yuan)
$u_{O\&M}$	Operation and maintenance cost (Yuan)
c_e	Economic cost (Yuan)
u_d	Unit environmental cost of carbon dioxide emission (Yuan/ton)
d	Unit carbon dioxide emission of energy consumption (g/kWh)
g	Energy consumption (kWh/year)
c_d	Environmental cost (Yuan/year)
u_p	Average increased public transport cost (Yuan/person · year)
p	Population density (persons /m^2)
l	Length of pipe construction (m)
c_s	Social cost (Yuan/year)
u_f	Unit cost saving on fertilizers (Yuan/m^3)
f	Amount of reclaimed water for agricultural irrigation (m^3/year)
b_f	Economic benefit (Yuan/year)
u_e	The monetary value of water (Yuan/m^3)
e	Amount of reclaimed water (m^3/year)
b_e	Environmental benefit (Yuan/year)
w	Number of plant workers (persons)
W	Total employment number in the region (persons)
Y	GDP of the region (Yuan)
B	Employment elasticity
b_s	Social benefit (Yuan)

3.3.3 Results of financial and economic analysis

Table 3.6 presents the results of the financial analysis of the Gao and the Jiu centralized wastewater reclamation plants. Table 3.6 shows that the initial investment of the Gao plant was 323.52 million Yuan and that of the Jiu plant was 76.96 million Yuan. As indicated in Tables 3.1 and 3.2, the treatment capacity of the Gao plant is 300,000 m^3 per day and the capacity of the Jiu plant is 60,000 m^3 per day. This means that the scale of the Gao plant is five times the scale of the Jiu plant. Accordingly the initial investment of the Gao plant is four times the investment of the Jiu plant. For centralized wastewater reuse systems, the initial investment depends on the capacity of the system.

Table 3.6
Financial analysis of centralized wastewater reuse systems

	Gao plant	**Jiu plant**
Financial cost		
Initial investment (million Yuan)		
Construction cost	166.31	38.33
Demolition and relocation cost	108.37	29.21
Preparation cost	11.77	5.68
Others	25.37	3.74
Sub-total	311.82	76.96
O&M cost (million Yuan/year)		
Energy cost	4.39	1.45
Chemical	1.8	0.44
Maintenance	3.48	1.29
Personnel	0.72	0.48
Interest rate	0.59	0
Sub-total	10.98	3.66
Financial Benefits		
Revenue (million Yuan/year)	576.96	20.3
Subsidies (million Yuan)	123.52	0

Table 3.6 shows that the construction cost and the demolition and relocation cost separately account for around 50 per cent and 35 per cent of total initial investment. The cost of construction and demolition and relocation are the main part of the initial investment of a centralized wastewater reuse system. As enormous investments are required for the construction of a centralized wastewater reclamation system, parts of that money may need to be borrowed, which means interest rate will need to be factored in the O&M cost. In this case, the Gao plant needs to pay for the interest rate of 0.59 million Yuan per year (Table 3.6) because it has a bank loan of 200 million Yuan (Beijing Water Authority 2002).

The energy cost is the highest cost in the O&M budget, accounting for around 40 per cent (shown in Table 3.6). Since centralized wastewater reuse systems use vast energy consumptions for water treatment, the electricity cost in China could have significant influence on the O&M cost of centralized systems. Besides the energy cost, maintenance cost is a large expense in the O&M budget. For instance the maintenance cost for the Gao plant was 3.48 million Yuan, which is 35 per cent of the O&M budget (shown in Table 3.6).

Table 3.7
Economic analysis of centralized wastewater reuse systems (million Yuan)

	Gao plant	**Jiu plant**
COST		
Economic cost	425.53	112.89
Environmental cost	19.24	0.38
Social cost	7.13	4.9
TOTAL	452	118
BENEFITS		
Economic benefits	2.4	0.24
Environmental benefits	3,225.26	645.05
Social benefits	5.9	5
TOTAL	3,233.6	650

The present value of all effects in the economic analysis are calculated and shown in Table 3.7. Because of the large treatment capacity of centralized wastewater reuse systems, a large quantity of reclaimed water is generated and reused, leading to environmental benefits of increasing water availability. In water scarce areas, water has a higher economic value. Thus the environmental benefits shown in Table 3.7 have a large value. The environmental cost of the Gao plant is 19.24 million Yuan whereas the environmental benefit of the Gao plant is 3,225.26 million Yuan. The environmental benefits are much higher than the environmental cost, which also occurs in the Jiu plant. Although there is a serious environmental cost caused by carbon dioxide emission, the environmental benefits are larger than the environmental cost.

About the social effects, the economic value of social cost is close to the value of the social benefits. The social cost is represented by residential resettlement due to pipe construction, and the social benefit is represented by the improvement in employment.

Table 3.8
Results of financial and economic feasibility of centralized wastewater reuse systems

	Gao plant	Jiu plant
Financial analysis (ratio of financial benefits to financial cost: $r_{fb/fc}$)	1.6	1.7
Economic analysis (ratio of benefits to cost: $r_{b/c}$)	7.2	5.5

Table 3.8 shows the results of the ratios of benefits (or financial benefits) to cost (or financial cost) in the economic and financial analysis. In the financial analysis, the ratios of financial benefits to financial cost of the Gao plant and the Jiu plant are 1.6 and 1.7 respectively, which are larger than 1. This means that the two plants are financially feasible. Given this situation, the plant managers have an incentive to operate the wastewater reuse plant. Although the construction cost and energy cost of a centralized wastewater reuse plant are very high, the revenue is high enough to

cover these costs. From the point of view of plant managers, centralized wastewater reuse systems can be operational in the long term.

In the economic analysis, the ratio of benefits to cost of the Gao plant is 7.2, which is larger than 1. Similarly the ratio of benefits to cost of the Jiu plant is larger than 1. This means that both the Gao plant and the Jiu plant are economically feasible. This implies that centralized wastewater reuse systems have a positive influence on the welfare of society. From the point of view of government, centralized wastewater reuse systems deserve to be promoted.

3.3.4 Conclusions of financial and economic analysis

The financial and economic analysis of centralized wastewater reuse systems in Beijing is carried out through cost benefit analysis. Similar to Chapter 2, the framework of this chapter includes two parts: financial analysis and economic analysis. The financial analysis is made from the point of view of plant manager, while the economic analysis is from the point of view of society. The major economic, environmental and social effects of the projects are all considered in the economic analysis.

The results of financial analysis show that the centralized wastewater reuse systems are financially feasible. This means that the investment on centralized wastewater reuse systems is profitable, which could raise the incentive of plant managers to operate the plant. From the point of view of plant manager, centralized wastewater reuse systems could operate in the long term.

The results of economic analysis show that the economic, social and environmental benefits caused by centralized wastewater reuse plants are larger than the cost. So the centralized wastewater reuse systems are economically feasible. This means centralized wastewater reuse systems have positive effects on society. From the point of view of government or society, the centralized wastewater reuse systems are worthy of promotion. Moreover, the results show that centralized wastewater reuse systems make a large positive contribution to the environment despite consuming substantial resources for construction and may increase carbon dioxide emissions.

In the literature, from the technological point of view, centralized wastewater reuse systems producing large quantities of new water could effectively reduce the pressure of urban water scarcity (Asano 2001). In terms of the results of this study, from an economic point of view, centralized wastewater reuse systems are a profitable investment and have a positive influence on the welfare of society.

3.4 Comparison between Decentralized and Centralized Wastewater Reuse Systems

Chapter 2 offered financial and economic analysis of decentralized wastewater reuse systems. Section 3.3 of this chapter implements the financial and economic analysis of centralized wastewater reuse systems. According to the results of Chapter 2 and section 3.3 of this chapter, a comparative analysis between decentralized and centralized wastewater reuse systems will now be carried out.

3.4.1 Financial and economic feasibility

Table 3.9
Comparison of financial and economic feasibility between decentralized and centralized wastewater reuse systems

	Decentralized		Centralized	
	Qing plant	BNU plant	Gao plant	Jiu plant
Capacity (m^3/day)	65	400	300,000	60,000
Financial ratio of benefits to cost	0.13	0.38	1.6	1.7
Financially feasible	No	No	Yes	Yes
Economical ratio of benefits to cost	4.7	5.4	7.2	5.5
Economically feasible	Yes	Yes	Yes	Yes

Table 3.9 illustrates that the economical ratios of benefits to cost of the centralized and decentralized plants are all larger than 1. From the viewpoint of government, both decentralized wastewater reuse systems and centralized wastewater reuse systems have positive effects on society because they are all economically feasible. Moreover, we find that the economic ratio of benefits to cost is higher when the capacity of the plant is

larger. For example, the Gao plant has the largest capacity and its economic ratio of benefits to cost is the highest. In contrast, the Qing plant with the smallest capacity has the lowest economic ratio of benefits to cost (shown in Table 3.9). Hence the large wastewater reuse systems have more significant positive effects on society.

In the financial analysis, only the ratios of the centralized wastewater reuse plants are larger than 1 (shown in Table 3.9). So the centralized wastewater reuse systems are financially feasible while the decentralized plants are not. In centralized systems, the financial cost could be recovered by the financial benefits, which ensures that the centralized wastewater reuse plants could operate continually. But in the decentralized wastewater reuse system, the financial benefit is not enough to recover the cost. So the plant managers would prefer not to operate the systems. It is a challenge to continue the operation of the decentralized wastewater reuse systems in the long term. Therefore, from the perspectives of plant managers, managing the centralized wastewater reuse systems is a better option than managing the decentralized systems.

In terms of the results of the economic and financial feasibility of the wastewater reuse in Beijing central region, centralized wastewater reuse systems are more economically and financially competitive than decentralized systems. The study of Jia et al. (2005) who carried out the analysis of the wastewater reuse in Beijing through the Geographic Information System (GIS) method also found that large wastewater reuse systems are more suitable for the Beijing central region.

3.4.2 Environmental and social effects

Table 3.10 presents the average environmental costs and benefits and the average social costs and benefits of the decentralized wastewater reuse systems and centralized wastewater reuse systems. The average cost (benefits) value is obtained by dividing the total cost (benefits) by the total volume of reclaimed water, which means how much environmental or social cost (benefits) can be caused by producing one cubic metre of reclaimed water.

Table 3.10 indicates that both the centralized and decentralized wastewater reuse systems have positive net environmental benefits because the environmental benefits are larger than the environmental costs.

The ratio of environmental benefits to cost indicates that in decentralized wastewater reuse systems, the net environmental benefits of the large plant are higher than the net benefits of the small plant, but in centralized systems, the net environmental benefits of the large plant is lower than the small plant.

Table 3.10
Comparison of environmental and social effects between decentralized and centralized wastewater reuse systems

	Decentralized		Centralized	
	Qing plant	BNU plant	Gao plant	Jiu plant
Capacity (m^3/day)	65	400	300,000	60,000
Environmental cost (Yuan/m^3)	0.14	0.007	0.009	0.0009
Environmental benefits (Yuan/m^3)	1.7	1.93	1.47	1.47
Ratio of environmental benefits to cost	12.35	259	167	1697
Social cost (Yuan/m^3)	0.05	0.009	0.003	0.01
Social benefits (Yuan/m^3)	0.09	0.2	0.003	0.011
Ratio of social benefits to cost	1.6	21	1	1.02

The effect of environmental benefit considered in the economic analysis of decentralized and centralized systems is the same, which is the benefit of increasing water availability. Table 3.10 shows that the average environmental benefits of decentralized systems are slightly higher than the environmental benefits of centralized systems. Moreover, the effects of environmental cost considered in the economic analysis of two systems are different: the environmental cost caused by decentralized systems is noise pollution while the environmental cost caused by centralized systems is carbon dioxide emission. The numbers in Table 3.10 illustrate that for decentralized systems, the large plants lead to less environmental cost than the small plants, while for centralized systems, the large plants leads to more environmental cost than the small plants.

The main social effects taken into account are different between decentralized wastewater reuse systems and centralized systems. For decentralized systems, the social cost is presented by health risk and the social benefit is presented by raising social awareness; while for centralized systems, the social cost is presented by the consequence of residential resettlement and the social benefit is presented by improved employment.

Table 3.10 shows that the average social benefits caused by decentralized and centralized wastewater reuse systems are larger or equal to the average social cost. So both decentralized and centralized systems have positive net social benefits. Moreover, for both decentralized and centralized systems, the larger plants cause less social cost. But the figures of social benefits reflect a different result: the large plants in decentralized systems lead to higher average social benefits while in centralized systems, the small plants lead to higher social benefits.

3.4.3 Initial investments

Table 3.11
Comparison of initial investments between decentralized and centralized wastewater reuse systems (million Yuan)

	Decentralized		Centralized	
	Qing plant	BNU plant	Gao plant	Jiu plant
Initial investment	2.9	3.7	323.5	77
Pipe construction	2.6	3.1	313	58
Demolition and relocation	0	0	108.4	29.2

Table 3.11 presents the initial investments including the cost of pipe construction of centralized and decentralized wastewater reuse plants. The initial investments of the centralized plants are much larger than the initial investment of the decentralized plants. For example, the investment of the Jiu plant is 76 million Yuan while the initial investment of the Qing plant is only 2.9 million Yuan (shown in Table 3.11). That is to say, the expenditure to construct one centralized plant (capacity: 60,000

m^3 per day) can be used for the construction of around 25 decentralized plants (capacity: 400 m^3 per day).

Table 3.11 shows that the cost of pipe construction is the major expenditure in the construction of wastewater reuse plants, accounting for around 80 per cent of the initial investments. Decentralized plants generally serve the surrounding areas, while the centralized plants serve the areas far away from the plants. The cost of pipe construction of the decentralized plants only contains the direct expenditure for pipe construction. But the cost of pipe construction of the centralized plants includes the expenditure of pipe construction, the cost of demolition and relocation of existing urban buildings or residences, and the interest rate.

The cost of demolition and relocation accounts for large part of the total cost of pipe construction in the centralized wastewater reuse systems. Indicated in Table 3.11, the cost of pipe construction of the Gao plant was 313 million Yuan in total, of which 108.4 million Yuan went to the cost of demolition and relocation. This huge investment for pipes sometimes will be financed through loans, leading to the cost of interest in the centralized wastewater reuse plants. For example, interest cost for the Gao plant was 11.7 million Yuan (shown in Table 3.6). Moreover, demolition and relocation of the existing building or residence may affect urban planning and sustainable urban development. However, in the decentralized systems, there was no cost for demolition and relocation and interest cost does not play a role.

3.4.4 About O&M cost

Four items are involved in the determination of O&M cost: energy, chemical, maintenance and personnel. Figures 3.3 and 3.4 illustrate the cost of these four items in centralized wastewater reuse plants and in decentralized plants. Apparently the highest column shown in Figures 3.3 and 3.4 is the energy cost. The energy cost accounts for 60 per cent of total O&M cost in the decentralized plants and 40 per cent of the O&M cost in the centralized plant. So the O&M cost of a wastewater reuse system is affected significantly by energy cost.

Figure 3.3
O&M cost per year for the Gao and Jiu plants

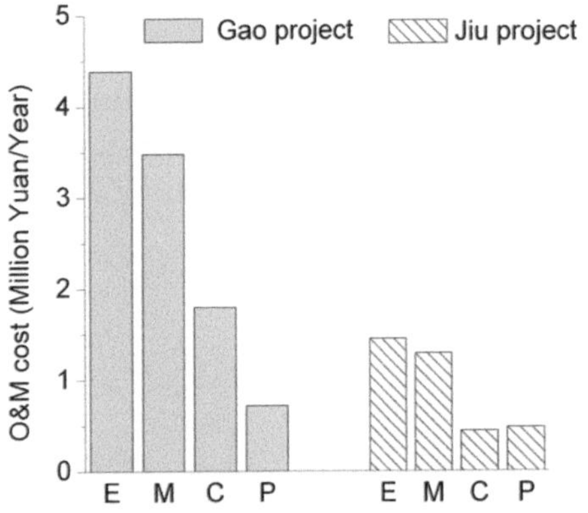

(E: energy, M: maintenance, C: chemical, P: personnel)

Figure 3.4
O&M cost per year for the Qing and BNU plants

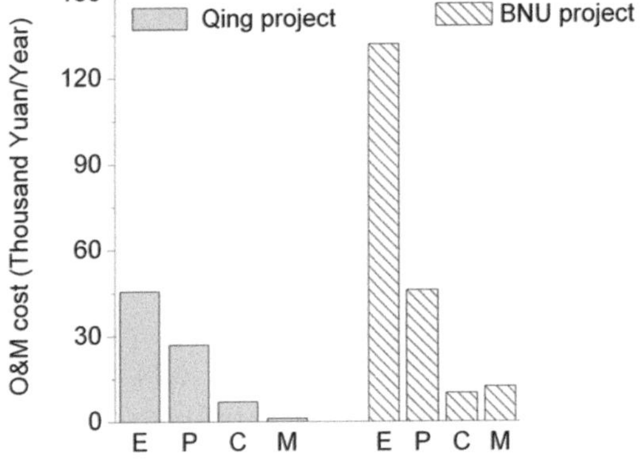

(E: energy, P: personnel, C: chemical, M: maintenance)

Figure 3.3 shows that the second highest column is the maintenance cost and there is still a small difference between the energy and maintenance cost. Since the treatment plant and the pipe network of a centralized wastewater reuse plant is quite large, the maintenance cost is an important part of the O&M cost. However, the second largest cost in decentralized systems is the personnel cost, which accounts for 20-30 per cent of the O&M cost (shown in Figure 3.4). Although the decentralized plant does not employ many workers, the salary payment for these employees is a high proportion of the O&M cost.

3.4.5 Cost recovery

Table 3.12
Comparison of cost recovery between decentralized and centralized wastewater reuse systems

	Decentralized		Centralized	
	Qing plant	BNU plant	Gao plant	Jiu plant
Capacity (m^3/day)	65	400	300,000	60,000
Unit O&M cost (Yuan/m^3)	3.8	1.5	0.09	0.16
Rate of O&M cost recovery	32%	0	565%	555%

The unit O&M cost shown in Table 3.12 is determined by dividing the O&M cost by the quantity of reclaimed water. In terms of scale economies, if a system has larger capacity, its unit O&M cost is much smaller. For example, the Gao plant being the largest system has the smallest unit O&M cost. Table 3.12 shows that the unit O&M cost of the Qing and BNU plants is 3.8 Yuan/m^3 and 1.2 Yuan/m^3 respectively, while the unit O&M cost of the Jiu and Gao plants is only 0.16 Yuan/m^3 and 0.09 Yuan/m^3. The unit O&M cost of the decentralized systems is almost 20 times more than the unit O&M cost of the centralized systems. The average cost of reclaiming wastewater in decentralized systems is more expensive than in centralized systems.

The revenue of a wastewater reuse plant depends on the rate of reclaimed water for different uses. Normally the decentralized systems are constructed for residential or industrial uses. The reclaimed water from centralized systems is mainly used for industrial cooling water, green irrigation and residential toilet flushing. Table 3.3 lists the rates of reclaimed water in terms of its utilization. Compared with the rates of reclaimed water for various utilizations, the unit O&M costs of the centralized plants is higher than all rates of reclaimed water for various utilizations while the unit O&M cost of the decentralized plants is lower than the rate of reclaimed water for different usages.

The rate of O&M cost recovery presented in Table 3.12 shows how much O&M cost is recovered by the revenue. Table 3.12 shows that the rate of O&M cost recovery of the BNU plant is zero because it does not charge for reclaimed water. The rate of cost recovery of the Qing plant is only 32 per cent, which is smaller than 1. In contrast, the rates of the Gao plant and the Jiu plant are around 500 per cent because the unit O&M cost of centralized plants is much higher than the rate of reclaimed water. The rate of cost recovery of the centralized wastewater reuse systems is higher than the rate of the decentralized systems.

The unit O&M cost of the decentralized systems, especially the small plant, is very high. The Qing plant was constructed for residential use and the charge for reclaimed water used for residences is 1.2 Yuan/m^3. But the unit O&M cost of the Qing plant is 3.8 Yuan/m^3, which is higher than the charge for reclaimed water used for residents. As a result the O&M cost of decentralized wastewater reuse systems could not be recovered. This may be improved if the rate of reclaimed water is increased. The current rates for reclaimed water are not suitable for decentralized systems, as the cost of decentralized systems cannot be recovered.

3.5 Conclusions

The economic and financial feasibility of the centralized wastewater reuse systems in Beijing is analysed in Chapter 3 according to the methodology shown in Figure 1.17 (Chapter 1). The financial analysis takes the point of view of plant managers while the economic analysis is from the point of view of society. The results show that the centralized wastewater reuse systems are not only financially feasible but also economically feasible. It implies that the centralized wastewater reuse systems in Beijing have positive effects on society and these systems could operate over the long-term.

Based on the results of Chapter 2 and the results of financial and economic analysis of centralized wastewater reuse systems, the economic and financial comparison between the centralized wastewater reuse systems and decentralized systems is performed from several perspectives. These are financial and economic feasibility, environmental and social effects, initial investment, O&M cost and cost recovery.

About the economic and financial feasibility, the centralized and decentralized wastewater reuse systems have different results. The decentralized plants are economically feasible but they are not financially feasible, while the centralized plants are both financially and economically feasible. Hence from the perspective of financial feasibility, the centralized wastewater reuse systems are more competitive than the decentralized systems.

The results of environmental comparison illustrate that both centralized and decentralized wastewater reuse systems have positive net environmental benefits because the environmental benefits are larger than the environmental cost. About the social effects, the centralized systems and decentralized systems lead to different effects, such as health risk and residential resettlement. The results of both systems illustrate that the social benefits are higher or close to the social cost.

Some findings are obtained through comparison of initial investments and O&M cost. The initial investments of the centralized systems are extremely big because of high expenditure on pipe construction. The cost of demolition and relocation and the interest paid for the loan are the main cost of pipe construction in centralized systems, but they rarely occur in decentralized systems. Moreover, for both centralized and decentralized wastewater reuse systems, the energy cost is the largest expense in the O&M cost. But the second largest expense in the O&M cost is different between two kinds of systems. The second largest expense in centralized systems is the maintenance cost while it is personnel cost in decentralized systems.

The difference between the unit O&M cost and the rate of reclaimed water leads to different implications for cost recovery. In centralized wastewater reuse systems, the unit O&M cost is lower than the rate of reclaimed water. But in decentralized systems, the unit O&M cost is higher than the rate of reclaimed water. Hence the rates of O&M cost recovery of centralized systems are much higher than the rate of cost recovery of decentralized systems. The current tariff of reclaimed water is not suitable for cost recovery of decentralized systems.

Notes

[2] Chapter 3 is submitted to the journal, *Journal of Benefit Cost Analysis*, 2010, under review.

4 Economic and Financial Analysis of Rainwater Harvesting Systems[3]

4.1 Introduction

As mentioned in Chapter 1, there are two main technological measures, wastewater reuse and rainwater harvesting, promoted in Chinese cities to prevent or solve urban water scarcity. Chapters 2 and 3 examined the wastewater reuse systems. The rainwater harvesting systems are studied in Chapters 4, 5 and 6.

Rainwater harvesting is an inexpensive and simple way to obtain water for agricultural irrigation. Using rainwater helps save water resources and solves the problem of environmental degradation (Li et al. 2000). Although hundreds of rainwater harvesting systems have been constructed in Beijing since 2006, most of these rainwater harvesting systems are concentrated in the rural areas of Beijing (Wang et al. 2007). The amount of rainwater collected in the urban areas of Beijing currently only accounts for ten per cent of the total rainwater collected in Beijing (Zuo et al. 2010). Because of the low volume of urban rainwater collection and because there are fewer rainwater harvesting projects in urban areas, limited data is available on urban rainwater harvesting. So the study focuses on rainwater harvesting systems in the rural areas of Beijing.

Rainwater harvesting systems in the rural areas of Beijing are used mainly for supplementing agricultural irrigation water. Continuous droughts and a decreasing groundwater table are causes for water scarcity for agricultural production in rural areas of Beijing. Both the amount of agricultural water consumption and the proportion of agricultural water consumption to total water consumption have decreased about 50 per cent during the last 20 years in Beijing including rural areas. Figure 4.1

shows agricultural water consumption in Beijing decreasing from 2.4 billion to 1.2 billion m^3, and the proportion of agricultural water to total water consumption in Beijing going down from 54 to 34 per cent. The decrease in consumption of agricultural water threatens the development of agricultural production, the ecological environment and the social stability of rural areas (Brown 2004; Wang and Wang 2005; Webber et al. 2008; Yang and Zehnder 2001). Rainwater harvesting in rural areas of Beijing could effectively supplement the supply of irrigation water.

Figure 4.1
Consumption and proportion of agricultural water

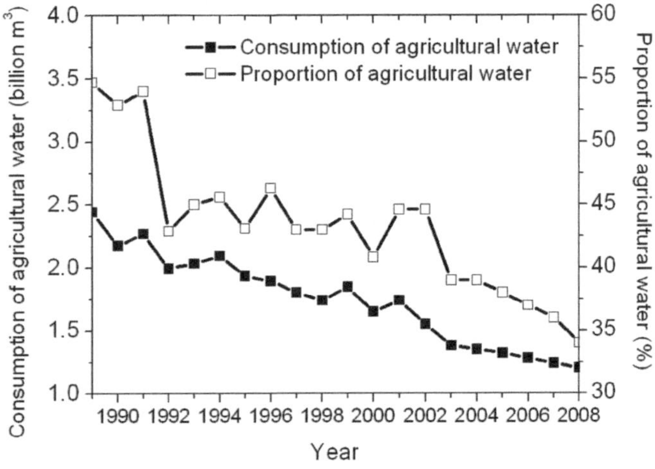

Source: Beijing Water Authority (1988-2009)

Considerable investments are required to build rainwater harvesting facilities to make irrigation water available for agricultural production. Hence it is important to understand the effectiveness of these investments. The rainwater harvesting systems in rural areas of Beijing are constructed because of the benefit of increased water resources, while the financial and economic implications of using these systems are discussed less. The project managers of these rainwater harvesting systems normally are local farmers who have no experience with or knowledge of rainwater harvesting. Compared with using groundwater, using rainwater for agricultural irrigation is new to local farmers. If the new systems were not more financially attractive than the original system, the farmers would reject them and continue to use ground water. Farmers' preference for water sources can determine whether the constructed rainwater harvesting systems will operate or not. Therefore the issue of whether using rainwater is financially more attractive than using groundwater is highly relevant.

The literature on economic analysis of rainwater harvesting mostly focuses on the evaluation of economical feasibility, based on different rainwater harvesting methods (Hatibu et al. 2006; Mushtaq et al. 2007; Pandey 1991; Tian et al. 2002). Few papers consider the effects of rainwater harvesting on the environment and society. One objective of this chapter is to do an economic analysis, quantifying the monetary values of the major economic, environmental and social effects caused by rainwater harvesting.

This chapter aims to analyse economic and financial performance of the rainwater harvesting systems in rural areas of Beijing through Cost Benefit Analysis (CBA). The economic analysis focuses on determining the contribution of rainwater harvesting systems to the development of society, carried out from the point of view of government. The financial analysis allows comparison of the financial implications of using groundwater with using rainwater for agricultural irrigation from the point of view of individual participant, namely the local farmers.

In this chapter, Section 4.2 introduces the data used for the analysis. Section 4.3 presents the methodology of the study, explaining how the economic and financial analysis is carried out. Sections 4.4 and 4.5 show the results and draw the conclusions.

4.2 About the Rainwater Harvesting Systems

The data for the estimation are average values, calculated based on Wang et al. (2007) and interviews with the project managers. Wang et al. (2007) research includes statistical data on Beijing rural rainwater harvesting. They work for the Beijing Agro-Technical Extension Center, which cooperates with the SWITCH programme. In addition, we carried out extensive interviews with managers of the rainwater harvesting projects.

Figure 4.2
Rainwater harvesting plant

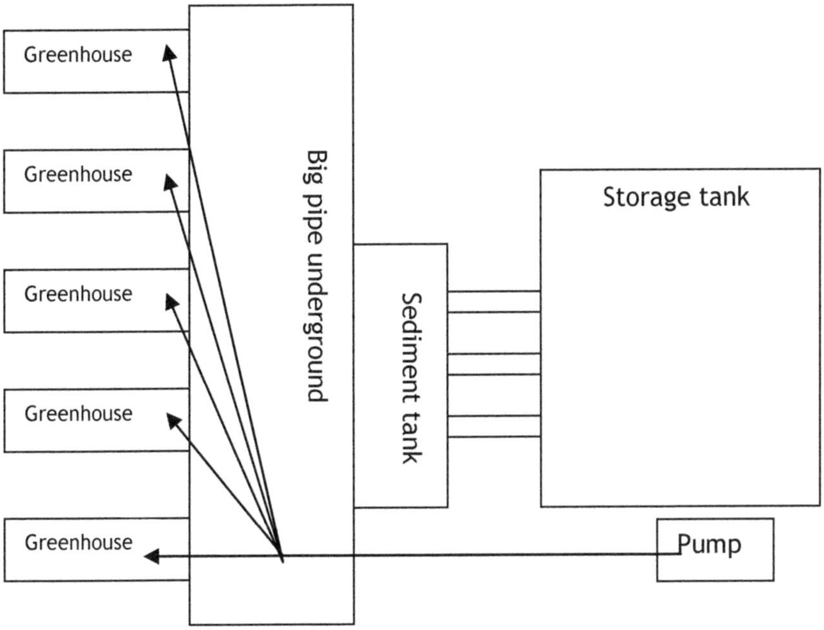

Around 600 rainwater harvesting systems were constructed for agricultural irrigation in Beijing in 2007 (Wang et al. 2007). Rainwater harvesting systems in rural areas consist of a rainwater collection part and a water reuse part. The rainwater collection part includes plastic covers for green houses, storage tanks and collective ditches, and the water reuse part is made of pumps and irrigation facilities. Figure 4.2 shows the structure of the An rainwater harvesting plant, which is a typical case. Most rainwater harvesting projects in rural areas of Beijing have a similar process for rainwater collection and reuse to the An plant. There is a plastic film covering the greenhouse, rainwater goes through the plastic film down to the ditch in front of the greenhouse. Rainwater moves from the shallow ditch to a big underground pipe and then to the sediment tank. There is a filter installed in the tank. After depositing the solids, cleaned water enters the storage tank. The water is pumped from the storage tank to the greenhouse. Normally the government subsidizes rainwater collection systems and the project managers pay the remaining initial costs.

The capacity of each rainwater harvesting tank ranges from about 50 to 1,000 m^3 and can be categorized into three sizes: small, medium and large rainwater harvesting systems. Our economic and financial analysis of this study relates to these three sizes. The relevant data are presented in Table 4.1.

Table 4.1
Data of rainwater harvesting systems with different sizes

Size	Capacity (m^3)	Average irrigation area (m^2/year)	Average irrigation amount (m^3/year)	Average users (person)
Small	50-100	4,002	1,200	8
Medium	450-800	53,360	16,000	80
Large	1,300-2,000	80,040	24,000	150

Source: Wang et al. (2007) and interviews with the project managers

Rainwater harvesting for agricultural irrigation in Beijing is currently in its infancy and suffers from technological problems and inefficient management. Hence less than the required amount of water for irrigation is harvested at present. It is expected that harvested rainwater can satisfy the required irrigation amount in the future (Wang *et al.*, 2007). In this study, we assume that the collected rainwater can satisfy the total amount of water required for agricultural irrigation.

4.3 Methodology

The methodology used for the financial and economic analysis of rainwater harvesting is Cost Benefit Analysis (CBA). The analytical framework of this chapter is based on the integrated financial and economic analysis method shown in Figure 1.17 (Chapter 1). Two parts of the evaluation framework are the economic and financial analysis. Figure 4.3 illustrates the evaluation framework of this chapter. The economic analysis is carried out through estimating economic, environmental and social effects. In the financial analysis, financial cost and financial benefits are evaluated. Additionally three different systems with different financial implications are compared. These three systems represent using groundwater (Option 1 and Option 2) and using rainwater (Option 3). The details of three options shown in Figure 4.3 represent Section 4.3.2. The aim of the comparison is to find which option is financially more attractive.

Figure 4.3
Two parts in the analysis

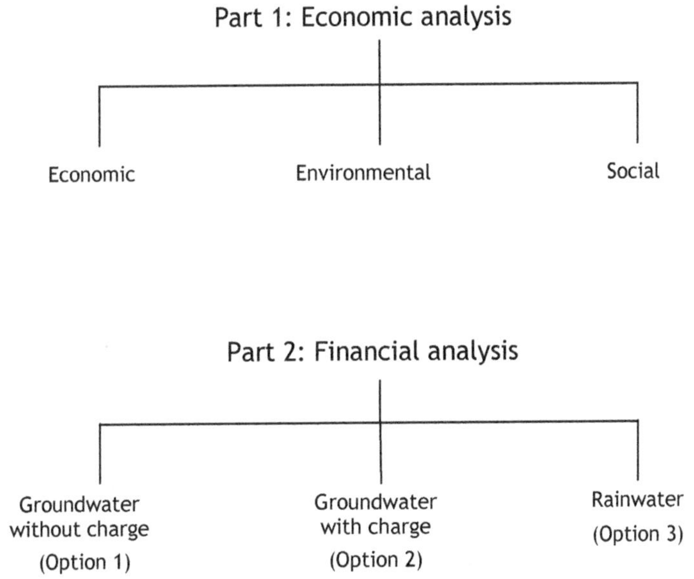

The CBA method evaluates the cost and benefits of economic, environmental and social effects in the economic analysis. The criterion of economic feasibility is the ratio of benefits to costs. If the ratio of benefits to costs is greater than 1, the rainwater harvesting project is economically feasible. If the ratio is less than 1, the project is not economically feasible.

In the financial analysis, the criterion of financial feasibility is whether using rainwater is financially attractive. Rainwater is the alternative to groundwater for agricultural irrigation. The ratio of financial benefits to financial costs or the difference between financial benefits and financial costs of rainwater harvesting is not enough to define whether the rainwater harvesting project is financially feasible or not. The farmers' preference for water resources, which are financially more attractive, determines financial feasibility of rainwater harvesting. If using rainwater is financially more attractive than using groundwater, the rainwater harvesting project is financially feasible. If using rainwater is not financially more attractive than using groundwater, the project is not financially feasible. The Net Present Value (NPV) for different water resources will be calculated. NPV is the difference between the present value of benefits and the present value of cost.

4.3.1 Economic analysis

Table 4.2
Economic, environmental and social effects of rainwater harvesting systems

Economic cost	Initial investment Operation and maintenance cost
Environmental cost	N*
Social cost	Agricultural risk
Economic benefit	Increase in agricultural production Planting in winter time
Environmental benefit	Water saving Energy saving
Social benefit	Raising social awareness Improving employment

*N= no significant environmental cost for rainwater harvesting systems.

In terms of the literature and interviews with officials at Beijing Water Authority and Beijing Agro-Technical Extension Center, the major economic, environmental and social effects caused by rainwater harvesting systems are listed in Table 4.2 (Tian et al. 2002; Wang et al. 2007). The process of collecting rainwater and then reusing it for agricultural irrigation is natural, so it is assumed that rainwater harvesting projects do not cause negative environmental effects. The environmental cost in Table 4.2 is 'N', which means no significant environmental cost. Similar to the economic analysis in Chapters 2 and 3, not all effects listed in Table 4.2 will be taken into account in the calculations. Only the main economic, environmental and social effects are selected and quantified using monetary values. The reasons for selecting effects and determination of values are explained below.

First, from a societal perspective, construction, and operation and maintenance (O&M) of rainwater harvesting systems consume scare resources, resulting in negative effects. Hence the initial investment and O&M costs are considered negative economic effects, calculated by Equations 4.1 and 4.2

$$E_C = C_N + C_{O\&M} \tag{4.1}$$

$$C_{O\&M} = \sum_{t=1}^{m} \frac{O_t}{(1+i)^t} \tag{4.2}$$

where E_C is the value of economic effects; C_N is the initial investment; $C_{O\&M}$ is the present value of O&M cost; O_t is the O&M cost occurring in year t, i is the discounting rate; m is the evaluation period (number of years). According to the publication *Chinese Economic Evaluation Parameters on Construction* (2006), the discount rate (i) used for cost benefits studies in China is eight per cent including the inflation rate. The evaluation period (m) is assumed to be ten years. As there are no traded items in the economic cost and there are no big distortions in market prices of water treatment construction in Beijing, the market prices could be used to determine monetary values of initial investment (C_I) and O& M cost (O_t).

Second, because many industrial plants are built around the city and domestic consumption creates significant domestic waste, air pollution is unavoidable in the city. Polluting chemicals in the air can pollute rainwater. When using rainwater for agricultural irrigation, the polluting chemicals enter agricultural plantings leading to serious agricultural risk. Agricultural risk due to using polluted rainwater could be regarded as a social cost caused by rainwater harvesting projects.

Equation 4.3 shows the social cost of agricultural risk (defined as E_A) could be determined by the average income from the vegetable (defined as a_i) and the total decreasing amount of agricultural production. In terms of interviews with project managers, it shows that cucumber, tomato, lettuce and marrow (a type of squash) comprise 80 per cent of crops cultivated through rainwater reuse systems. In the study, only the production of these crops is considered in the determination. According to the study by Wang et al. (2007), average incomes from these vegetables in Beijing (a_i) are around 2.5 Yuan per kilo ground (kg). The amount of decrease in agricultural production can be determined by multiplying the vegetable production and the decreasing rate of vegetable production due to polluted irrigation water (defined as a_r). Due to use of polluted water for agricultural irrigation, agricultural production can decrease by ten per cent (a_r) each year (Kang and Meng 1994; Kuang and Sun 1998). Then, the vegetable production can be obtained by the unit agricultural production (defined as a_p) and the irrigation area (defined as a_s). In the *Beijing Statistic Yearbook* (2010), the average vegetable production (a_p) in Beijing is 4.6 kg per square metres. The irrigation area of the rainwater harvesting systems (a_s) in the study is shown in Table 4.1.

$$E_A = a_i \times \left(a_p \times a_s \times a_r \times 80\%\right) \tag{4.3}$$

Third, because it is assumed that the volume of collected rainwater only satisfies the irrigation requirement, there is no increase in the volume of irrigation water leading to increasing agricultural production. Hence the economic benefit of increasing agricultural production is not considered in this study. Moreover, generally the storage tank is idle in the wintertime because the rainfall in Beijing is concentrated in the period from March to September. In some cases, the storage tank is used as a planted area in winter. The storage tank is a good place for shade-requiring plants, such as mushrooms. It is assumed in the study that all storage tanks in rainwater harvesting systems are used for planting in winter. The income from the planting could be regarded as an economic benefit of rainwater harvesting systems. According to interviews with project managers, mushroom planting in winter has been implemented in some areas of Beijing. To simplify the study, we assume that all rainwater harvesting systems plant mushrooms in winter and obtain income from it. Hence the economic benefit (defined as E_P) is determined by the unit income from mushrooms (defined as a_m) and the area of storage tanks (defined as S_P).

$$E_P = a_m \times S_P \tag{4.4}$$

Fourth, rainwater harvesting systems benefit by saving not only groundwater resources but also energy consumption. Table 4.3 distinguishes the traditional system of using groundwater (Case A) from the new system of using rainwater (Case B) for agricultural irrigation to clarify how much groundwater can be saved. It assumes that the harvested rainwater is X m^3 and the irrigation requirement is also X m^3. Case A represents the traditional system of using groundwater for irrigation, in which rainwater is not harvested and the precipitation cannot irrigate the crops directly since the crops are mostly planted in greenhouses. Hence only groundwater is used for irrigation in Case A. According to the *Beijing Water Resources Bulletin* (1988-1998), 10 per cent of the precipitation in rural areas of Beijing is recharged into groundwater and the remaining 90 per cent of the precipitation evaporates or becomes run off. Thus, in Case A, 10 per cent X m^3 rainwater penetrates the soil, while X m^3 groundwater is used for irrigation. Conversely, in Case B, X m^3 rainwater is harvested and reused for agricultural irrigation with no groundwater usage. When irrigating the crops, the amount of irrigation water penetrating the soil in Case A is the same as in Case B, which is assumed to be Y m^3. Comparing Case A with Case B, it can be concluded that the groundwater saving is 90 per cent X m^3 (Table 4.3). It implies that rainwater harvesting systems can save 90 per cent X m^3 groundwater when harvesting X m^3 rainwater.

Table 4.3
Groundwater saving amount

Systems	Sources	Use for irrigation	Groundwater resources depletion	Groundwater saving amount
Case A: Traditional system	a. Groundwater: pumping X m^3 b. Rainwater: No rainwater is harvested. 10% X m^3 of the rain-water penetrates the soil.	a. X m^3 groundwater b. 0 m^3 rainwater	X m^3 pumping groundwater -10%X m^3 rainwater pene-tration -Y m^3 irrigation water penetration	= Case A - Case B = X -10%X -Y-(-Y) = 90% X m^3
Case B: Rainwater harvesting system	a. Groundwater: no pumping b. Rainwater: All X m^3 rainwater is harvested. No rainwater pene-trates the soil.	a. 0 m^3 groundwater b. X m^3 rainwater	- Y m^3 irrigation water penetration	

Moreover, taking water from rainwater harvesting tanks consumes less energy than pumping water from underground. The energy consumption depends on the depth of the water to be pumped. Normally the depth of a well ranges from 80 to 100 metres while the depth of a tank ranges from two to eight metres. If there were no rainwater harvesting system, it would take 0.27 KWh/m^3 of energy consumption for irrigation; while if there were a rainwater harvesting system, energy consumption would decrease to 0.02 KWh/m^3. Consequently the unit energy saving is 0.25 KWh/m^3. Therefore, both the benefits of water and energy savings are regarded as environmental effects, determined by Equations 4.5, 4.6 and 4.7

$$E_B = E_W + E_N \tag{4.5}$$

$$E_W = U_W \times W_S \tag{4.6}$$

$$E_N = U_N \times N \times W_R \tag{4.7}$$

where E_B is the value of environmental effects; E_W is the value of water saving; E_N is the value of energy saving; U_W is the unit value of water; U_N is the unit value of energy; N is the unit energy saving; W_S is the amount of saved groundwater, which is 90 per cent of W_R; W_R is the amount of rainwater harvesting.

About the unit value of water (U_W) and the unit value of energy (U_N), we use the values from the literature directly. Chen et al. (2006) estimate the monetary value of groundwater in China is 6.5 Yuan/m^3. The value of 6.5 Yuan/m^3 can be used as the unit value of groundwater (U_W).

Since 70 per cent of Chinese energy generation is from coal power, the cost of coal power plants in China is regarded as the value of energy in this study (Dahowski et al. 2009). Skeer and Wang (2006) estimate the total cost of a coal power plant in China, including capital, fuel, O&M and pollutant. The result of this study shows that the base cost of coal power plant in China is $0.03 per kWh (namely 0.2 Yuan per kWh).

Finally, raising general awareness of water saving is reached through many different public education and advertisement campaigns. The introduction of rainwater harvesting systems is a method to enhance awareness and save costs on future awareness raising campaigns. We assume that the educational effect due to using rainwater harvesting systems is the same as the effect of public campaigns. Thus the cost savings on campaigns can be regarded as a social benefit of rainwater harvesting systems. As mentioned previously, the total expenditure on public awareness raising campaigns for water saving (defined as K) in Beijing is about 2.78 million Yuan/year, and around 2.25 million people (defined as M) are influenced by the campaigns (DPP 2001). The number of users affected by the rainwater harvesting systems (defined as E_u) is indicated in Table 4.1. Equation 4.8 determines the value of social effects of raising social awareness (defined as E_{aw}).

$$E_{aw} = E_u \times \frac{K}{M} \tag{4.8}$$

For the medium- and large-scale rainwater harvesting plants, some workers are required to operate and maintain the plant. This generates employment in the region. Fast economic growth can also help to improve employment. We assume that the employment effect of a large rainwater harvesting system is the same as the effect of economic growth. That means the value of economic growth contributing to improving employment can be regarded as a benefit of improving employment through rainwater harvesting. But small rainwater harvesting systems do not bring the benefit of improved employment. Typically small plants are operated and maintained by plant owners. The small plants do not need other workers to work on it. Hence the social benefit of improving employment is only considered in the medium and large rainwater harvesting systems. The determination of social benefit of improving employment caused by rainwater harvesting is shown in Equation 4.9. According to the definition in economics, the employment elasticity (defined as β) equals the ratio of the increase of employment (defined as $\frac{E_w}{W}$) to the increase of economic growth (defined as $\frac{E_{em}}{Y}$), in which E_w is the number of workers of the rainwater harvesting plant, W is the total employment of the region, Y is the GDP of the region, E_{em} is the value of economic growth namely social benefit of employment improvement. The employment elasticity of China is estimated to be 0.3, which means an increase in economic growth by one per cent can increase employment by 0.3 per cent (Li 2003). Moreover, a medium size rainwater harvesting plant generally requires one worker to manage the plant, and a large size plant needs two workers. So the social benefits of medium and large size rainwater harvesting systems are calculated by adding the benefit of increasing social awareness (Equation 4.8) with the benefit of improving employment (Equation 4.9), while the social benefits of small size rainwater harvesting systems only consider the benefit of increasing social awareness (Equation 4.8).

$$E_{em} = \frac{\frac{E_w}{W}}{\beta} \times Y \tag{4.9}$$

After quantifying the economic, environmental and social costs and benefits, the present value is calculated. Equation 4.10 determines the present value of costs (defined as EC_{PV}), Equation 4.11 calculates the present value of benefits for the small size rainwater harvesting systems (defined as EB_{PV1}) and Equation 4.12 calculates the present value of benefits for the medium and large-size systems (defined as EB_{PV2}).

The ratio of benefits to cost is the criterion for economic feasibility, which is determined by Equations 4.13 and 4.14. The ratio of benefits to costs for the small-size rainwater harvesting systems (defined as $Er1_{b/c}$) is calculated by Equation 4.13, while the ratio of benefits to costs for the medium and large systems (defined as $Er2_{b/c}$) is calculated by Equation 4.14.

$$EC_{PV} = \sum_{t=1}^{m} \frac{E_A}{(1+i)^t} + C \tag{4.10}$$

$$EB_{PV1} = \sum_{t=1}^{m} \frac{E_P}{(1+i)^t} + \sum_{t=1}^{m} \frac{E_B}{(1+i)^t} + \sum_{t=1}^{m} \frac{E_{aw}}{(1+i)^t} \tag{4.11}$$

$$EB_{PV2} = \sum_{t=1}^{m} \frac{E_P}{(1+i)^t} + \sum_{t=1}^{m} \frac{E_B}{(1+i)^t} + \sum_{t=1}^{m} \frac{E_{aw}}{(1+i)^t} + E_{em} \tag{4.12}$$

$$Er1_{b/c} = \frac{EB_{PV1}}{EC_{PV}} \tag{4.13}$$

$$Er2_{b/c} = \frac{EB_{PV2}}{EC_{PV}} \tag{4.14}$$

Table 4.4

Summary of the parameters used for determination of cost and benefits of rainwater harvesting systems

Parameter	Definition
C_N	Initial investment (Yuan)
$C_{O\&M}$	Present value of O&M cost (Yuan)
O_t	O&M cost occurring in year t (Yuan/year)
E_C	Economic cost (Yuan)
a_i	Average income from the vegetable (Yuan/kg)
a_p	Unit agricultural production (kg/m^2)
a_s	Irrigation area (m^2)
a_r	Decreasing rate of vegetable production due to polluted irrigation water (%)
E_A	Social cost (Yuan/year)
a_m	Unit income from mushrooms (Yuan/m^2. year)
S_P	Area of storage tanks (m^2)
E_P	Economic benefit (Yuan/year)
U_W	Unit value of water (Yuan/m^3)
W_S	Amount of saved groundwater (m^3/year)
E_W	Value of water saving (Yuan/year)
U_N	Unit value of energy (Yuan/KWh)
N	Unit energy saving (KWh/m^3)
W_R	Amount of rainwater harvesting (m^3/year)
E_N	Value of energy saving (Yuan/year)
E_B	Environmental benefit (Yuan/year)
K	Total expenditure on public awareness raising campaigns (Yuan/year)
M	Population affected by public campaigns (persons)
E_u	Number of users affected by the rainwater harvesting systems (persons)
E_{aw}	Social benefit of increasing social awareness (Yuan/year)
β	Employment elasticity
W	Total employment of the region (persons)
Y	The GDP of the region
E_w	The number of workers at the rainwater harvesting plant (persons)
E_{em}	Social benefit of employment improvement (Yuan)

All the parameters used to determine the monetary values of economic, environmental and social effects are summarized in Table 4.4.

4.3.2 Financial analysis

In this case, using groundwater represents the 'original system' and using rainwater represents the 'new system'. The purpose of comparing the financial implications of using groundwater with using rainwater is to find which one is financially more attractive system, and which factors determine the effectiveness of investment in rainwater harvesting systems. The municipal government is going to implement a charge for groundwater, which may change the financial implications of using groundwater. Hence three options for using different water resources for agricultural irrigation are specified:

Option 1: using groundwater for agricultural irrigation without charge. As there is no rainwater harvesting system, the irrigation water is all pumped from the well.

Option 2: using groundwater for agricultural irrigation while paying a water charge. As there is no rainwater harvesting system, the irrigation water is all pumped from the well. The groundwater is supposed to be charged by the government at 2 Yuan/m^3.

Option 3: using rainwater for agricultural irrigation. All irrigation water comes from rainwater due to the rainwater harvesting systems. The government subsidizes the rainwater harvesting take, but the project manager should pay the remaining costs.

The Net Present Value (NPV) of these three options is calculated by Equations 4.16, 4.17 and 4.18. Financial cost and benefits are two principal factors determining the NPV. Financial cost includes initial investment (C_I) and O&M cost (O_t), and financial benefits include subsidies and the income from four kinds of main crops: cucumber, tomato, lettuce and marrow. In financial analysis, market price is used to value the financial cost and benefits.

$$B_n = \sum_{t=1}^{n} \frac{I_t + S_t}{(1+i)^t} \tag{4.16}$$

$$C_n = C_I + \sum_{t=1}^{n} \frac{O_t}{(1+i)^t} \tag{4.17}$$

$$V_n = B_n - C_n \tag{4.18}$$

where n is the evaluation period (number of years); B_n is the present value of the financial benefits in the evaluation period of n; I_t is the income from crops in year t; S_t is the subsidies occurring in year t; C_n is the present value of the financial cost in the evaluation period of n; V_n is the NPV in the evaluation period of n.

In this study, the evaluation period n is a set of values ranging from 1 to 10. Hence there are ten net present values, namely, V_1, V_2, V_3...V_{10}. This approach of calculating NPVs at different evaluation periods facilitates the discovery of capital recovery periods and comparison of the financial performances of different options.

As mentioned, the criterion of financial feasibility in this chapter is whether using rainwater is financially more attractive than using groundwater, which can be determined by comparing three options of using different water resources for agricultural irrigation. If using rainwater is financially more attractive than using groundwater, the rainwater harvesting project is financially feasible. If using rainwater is not financially more attractive than using groundwater, the project is not financially feasible.

4.4 Results of Economic Analysis

Table 4.5 shows the results of economic analysis of the three sizes of rainwater harvesting plants, in which the present values of all effects are calculated and listed. The ratio of benefits to costs is the criterion of economic feasibility. Table 4.5 indicates that the ratio of benefit to cost for a small plant is 1.6, the ratio for a medium plant is 2 and the ratio for a large plant is 2.5. The ratios of three sizes of rainwater harvesting systems are all greater than 1, which means the rainwater harvesting systems are economically feasible. The rainwater harvesting systems have positive effects on the society. From the point of view of government, constructing rainwater harvesting systems should be promoted.

Table 4.5
Economic analysis of rainwater harvesting systems (Yuan)

	Small size systems	Medium size systems	Large size systems
COST			
Economic cost	25,403	278,314	424,719
Social cost	24,705	329,406	494,109
TOTAL	50,108	607,720	918,828
BENEFITS			
Economic benefits	26,840	241,563	697,848
Environmental benefits	52,983	706,437	1,059,656
Social benefits	66	285,773	571,465
TOTAL	79,889	1,233,773	2,328,969
RATIO (Benefits/Cost)	1.6	2	2.5

As mentioned, the social benefits of small plants include only the benefit of raising social awareness while the medium and large plants contain the benefit of raising social awareness and increase in employment. Hence in Table 4.5, the value of the social benefits of small plants is much smaller than the social benefits of medium and large plants.

Table 4.5 shows that the environmental benefits of small plants are 52,983 Yuan, whereas the economic cost and social cost of small plants is 25,403 and 24,705 Yuan separately. The environmental benefits are almost twice that of the economic cost or the social cost. Similarly for medium and large plants, the remarkable values of environmental benefits are much larger than the value of costs. This implies that rainwater harvesting systems make a positive contribution to the environment despite consuming resources during construction and the potential for agricultural risk.

4.5 Results of Financial Analysis

4.5.1 Comparison between the three options

Table 4.6
List of financial cost and benefits of different system sizes

	Option 1	Option 2	Option 3
Small size systems			
Financial cost			
Initial investment (Yuan)	0	0	27,000
O&M Cost (Yuan/year)	156	2,556	60
Financial benefits			
Subsidies (Yuan)	0	0	9,000
Income from crops (Yuan/year)	5,000	5,000	5,000
Medium size systems			
Financial cost			
Initial investment (Yuan)	0	0	300,000
O&M Cost (Yuan/year)	2,080	34,080	800
Financial benefits			
Subsidies (Yuan)	0	0	150,000
Income from crops (Yuan/year)	98,000	98,000	98,000
Large size systems			
Financial cost			
Initial investment (Yuan)	0	0	450,000
O&M Cost (Yuan/year)	3,120	51,120	1,200
Financial benefits			
Subsidies (Yuan)	0	0	300,000
Income from crops (Yuan/year)	137,000	137,000	137,000

Table 4.6 lists the data used in financial analysis, in which options 1 and 2 represent using groundwater and option 3 represents using rainwater. As the groundwater for agricultural irrigation is mainly taken from a well, dug long ago, privately or by village collectives, the initial cost to farmers of using groundwater is almost zero, which is reflected in Table 4.6. In contrast, using rainwater requires significant initial investment. However, the O&M cost of using rainwater is much less than using groundwater due to reduced pumping cost. The increase in well depth due to the decline of groundwater level leads to higher cost for pumping groundwater. But, less cost is required to pump rainwater from a storage tank to irrigation. Hence using groundwater and using rainwater have their own financial advantages and disadvantages, leading to significant financial implications.

Figure 4.4
Comparison between options 1, 2 and 3 of small-size systems

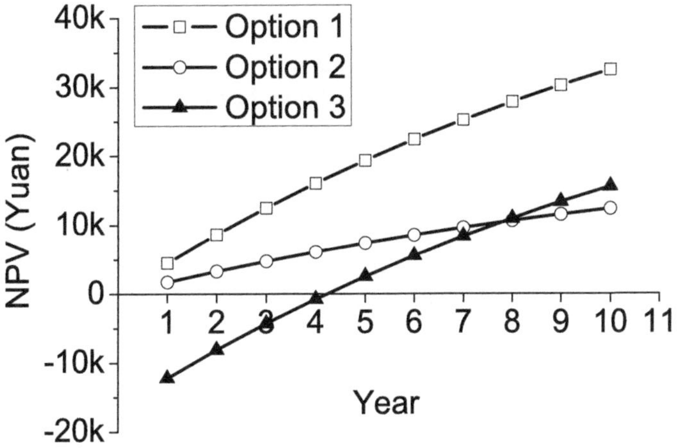

Figure 4.5
Comparison between options 1, 2 and 3 of medium-size systems

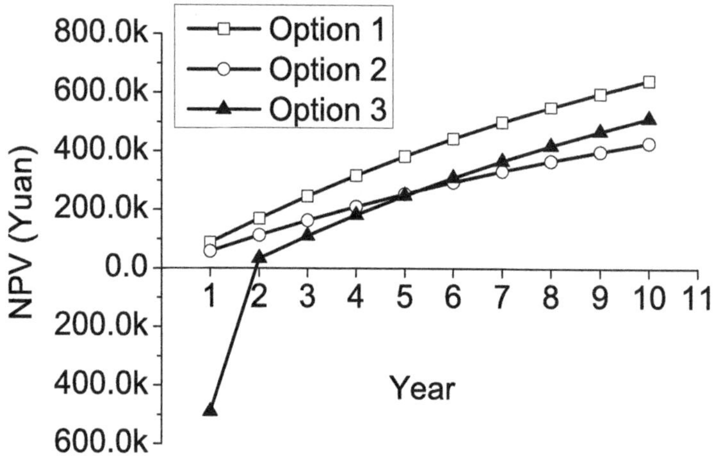

Figure 4.6
Comparison between options 1, 2 and 3 of large-size systems

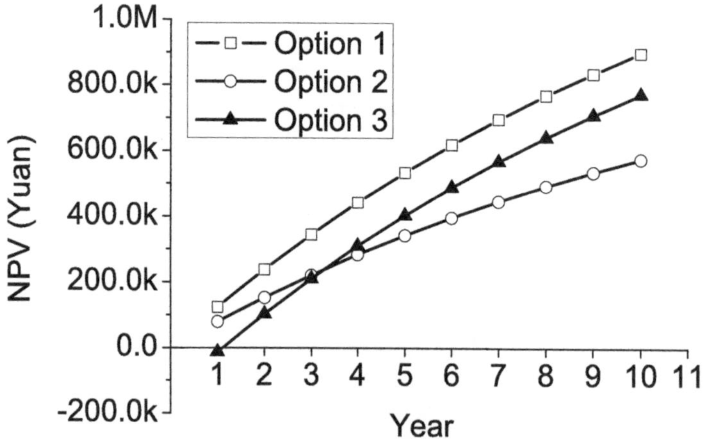

Figures 4.4, 4.5 and 4.6 separately show the comparison of the NPV lines of three options in three sizes of rainwater harvesting systems. Through comparing the NPV lines of options 1 and 3, we find that if groundwater is not charged, using rainwater is financially less attractive than using groundwater. It is shown in Figures 4.4, 4.5 and 4.6 that the NPV lines of option 3 of three sizes of systems are lower than the NPV lines of option 1 because using groundwater without charge (option 1) affords less cost than using rainwater (option 3). Given this situation, farmers would prefer to use groundwater for agricultural irrigation. Accordingly the constructed rainwater harvesting systems will not be used if groundwater is not charged.

On the other hand, if groundwater is charged at 2 Yuan/m^3, the situation will become complicated. The analysis of comparing the NPV lines of options 2 and 3 is as follows. In small-size systems (Figure 4.4), option 3 would take eight years to become financially better than option 2, and in medium-size systems (Figure 4.5), option 3 would take only five years to become better than option 2. However, in large size systems (Figure 4.6), option 3 would take only three years to become better than option 2. Furthermore, the capital recovery period of large-size systems is shorter than the capital recovery periods of small and medium-size systems. As illustrated in Figures 4.4, 4.5 and 4.6, the capital recovery periods of small, medium and large-size systems are respectively four years, two years and one year. Taking long periods to recover capital investment and to overtake the original system could weaken farmers' incentive to invest in, or to operate the new system. Of the three sizes of systems, small and medium-size systems take too long to recover capital costs and financially overtake using groundwater, but large-size systems only take one year for capital cost recovery and three years to overtake. The graphic analysis demonstrates that when groundwater is charged, using rainwater in large-size systems is financially more attractive than using groundwater, but in small and medium-size systems it is not financially attractive. Given this situation, only large systems can work smoothly, while farmers may not use small and medium systems.

The discussion above illustrates that the financial feasibility of rainwater harvesting systems depend on charging for groundwater. If there is no charge for groundwater, none of the three sizes of rainwater harvesting plants is financially feasible. If groundwater is charged at 2 Yuan/m^3, only large plants are financially feasible while small and medium plants are not financially feasible.

4.5.2 Further discussion

Previous sections demonstrate that using rainwater in small or medium-size systems is less financially attractive than using groundwater no matter whether or not they are charged for groundwater. But financial performance could be enhanced through decreasing financial costs or increasing financial benefits. Decreasing financial costs can be achieved through reducing the size of rainwater harvesting tanks. In this case, the government constructed the tanks so the size of tanks cannot be changed. Thus decreasing financial cost is not discussed in this study. Increasing financial benefits, as another way to improve financial performance, can be achieved through increasing subsidies or increasing income from crops.

These two factors are added in the financial analysis of small and medium-size systems for further discussion in this study. Thus two alternatives, option 3(a) and option 3(b) are presented:

Option 3(a): Increasing subsidies: Not only the harvesting tank but also half the remaining cost of initial investment is subsidized.

Option 3(b): Increasing income from crops: There is an increase in income from crops. We assume that the harvesting tank is used for mushroom planting when not used for water storage.

The results in Figures 4.7 and 4.8 illustrate that in small and medium-size systems, the NPV values of option 3(a) and 3(b) are both higher than that of option 3. It proves increasing subsidies or increasing income from crops can improve financial performance.

Figure 4.7
Comparison between different options of small-size systems

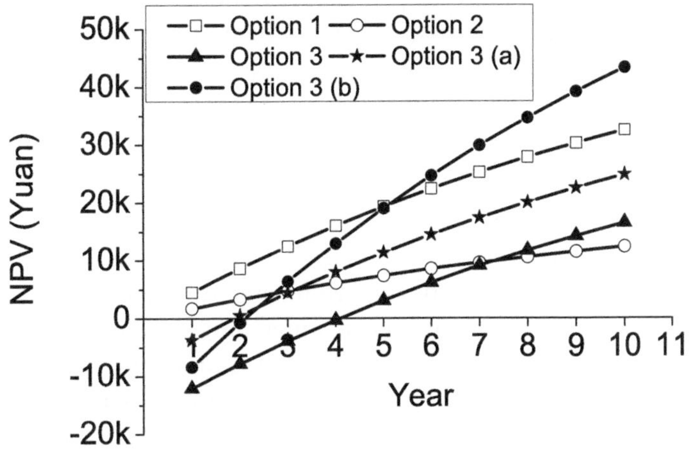

Figure 4.8
Comparison between different options of medium-size systems

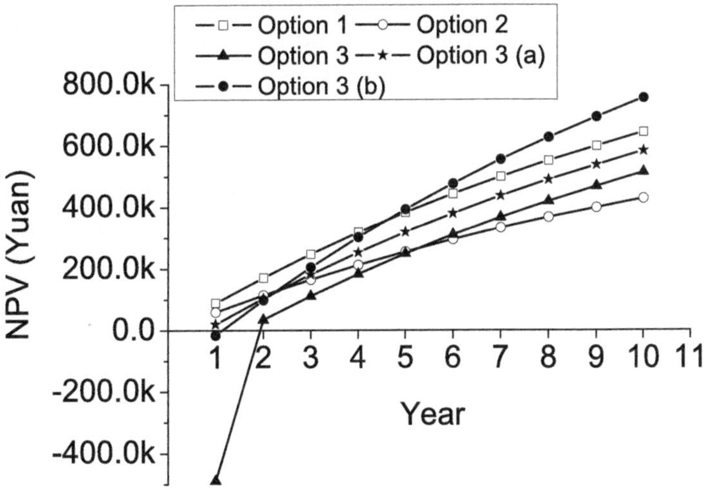

For both small and medium-size systems (Figures 4.7 and 4.8), the NPVs of option 3(a) and 3(b) would take only three years to become higher that the NPV of option 2, and the capital recovery periods become shorter. That means increasing financial benefits can lead to using rainwater in small and medium-size systems being financially more attractive than using groundwater when groundwater is charged. Figures 4.7 and 4.8 illustrate that when groundwater is not charged; small and medium-size systems are still financially less attractive than using groundwater. As such, the NPV of option 3(a) does not overtake the NPV of option 1, and option 3(b) takes a long time to overtake option 1.

Comparing option 3(a) and option 3(b), we find that option 3(b) is financially better than option 3(a). Referring to Figures 4.7 and 4.8, although option 3(b) requires more initial investment than option 3(a) does, the capital recovery periods of both options are the same and the NPV of option 3(b) becomes higher than option 3(a) after two years. This implies increasing income is a more effective factor than increasing subsidies to improve financial performance of small and medium-size systems.

4.6 Conclusions

This chapter presents an economic and financial analysis of rainwater harvesting systems in Beijing. An extended economic analysis evaluates the major economic, environmental and social effects of rainwater harvesting systems, while a financial analysis focuses on comparing the financial implications of using groundwater with using rainwater for agricultural irrigation.

The results of the economic analysis show that the small, medium and large sizes of rainwater harvesting systems are all economically feasible. This means that rainwater harvesting systems have positive effects on society. From the point of view of the government, rainwater harvesting systems are worth being promoted.

However, the results of the financial analysis indicate that the financial feasibility of rainwater harvesting systems depends on the charge for groundwater. If groundwater is not charged, three sizes of rainwater harvesting systems are not financially feasible because using rainwater is financially less attractive than using groundwater. Given this situation, farmers would continue to use groundwater for agricultural irrigation. If groundwater is charged at two Yuan/m^3, only large-size systems are financially feasible while small and medium-size systems are not financially feasible. Under these circumstances, only large systems can run smoothly, while farmers may not use the small and medium-size systems. Further financial analysis demonstrates that financial performance of small and medium-size systems can be improved through increasing subsidies or increasing income from crops.

This chapter finds that the financial feasibility of rainwater harvesting systems relates to groundwater charge. A charge imposed on groundwater raises the cost of groundwater. Comparatively rainwater harvesting becomes a more attractive option thereby improving the financial feasibility of rainwater harvesting. But imposing a charge on groundwater may discourage farming. How to improve the financial feasibility of rainwater harvesting through charging for groundwater is studied in the next chapter.

Notes

[3] Chapter 4 is submitted to the journal, *Resources, Conservation and Recycling*, 2010, under review.

5 Groundwater Charge and Rainwater Consumption for Agricultural Water Management[4]

5.1 Introduction

In Chapter 4, the results show that the financial feasibility of rainwater harvesting depends on charging for groundwater. If groundwater is not charged, no rainwater harvesting plants, regardless of size are financially feasible. If groundwater is charged at 2 Yuan per m^3, only large rainwater harvesting plants are financially feasible while small and medium size plants are not. If the charge for groundwater increases, rainwater harvesting could become financially more attractive, but it also could discourage farming. In Chapter 5, the relationship between groundwater charges and rainwater use is studied.

Groundwater contributes 75.6 per cent of the irrigation water for agriculture in Beijing (Wang and Wang 2005). Recently the groundwater stock in Beijing decreased drastically, which forced the government to stimulate rainwater harvesting to supplement water for irrigation.

The cost of using groundwater in Beijing is currently relatively low. Data in Table 4.7 (Chapter 4) illustrates that although the O&M cost of using rainwater is lower than the O&M cost of using groundwater, the initial investment cost of using rainwater is much higher than the initial investment cost for using groundwater. The government subsidizes only parts of the initial investment for rainwater harvesting, with the remaining amount paid for by farmers. The financial analysis in Chapter 4 shows that it takes three-to-five years to recover this initial investment in rainwater harvesting. But there is no initial investment for using groundwater. As the groundwater for agricultural irrigation is mainly taken from wells dug a long time ago, privately or by village collectives, the initial investment of digging wells is almost zero. Pumping and maintenance expenditures are the only other costs of using groundwater. At present, there is no charge for the use of groundwater in Beijing. Therefore, compared with rainwater harvesting, the cost of obtaining groundwater is relatively low.

Due to the relatively low cost for using groundwater, farmers have few incentives to use rainwater. Chapter 1 states that only 30 per cent of the constructed rainwater harvesting systems is used. From the point of view of technology and environmental effects, rainwater is a suitable replacement for groundwater in agriculture (Boers and Ben-Asher 1982; Li et al. 2000; Partzsch 2009). However, from a financial point of view, using rainwater is actually more expensive than using groundwater.

According to the economic theory of substitution, an increase in the price of one good will lead to an increase in demand for its substitute. It means that an increase in the cost of groundwater can raise the consumption of rainwater. If the cost of using groundwater were higher than using rainwater, farmers would prefer to reduce the consumption of groundwater while increasing the consumption of rainwater. Enforcing a levy on groundwater could be one method to increase the cost and restrict the consumption of groundwater (Webber et al. 2008). The Beijing Water Authority is the organization that should raise the cost for using groundwater by collecting a fee for groundwater use.

Higher cost for groundwater can effectively reduce groundwater consumption and motivate farmers to use rainwater. Meanwhile higher cost for groundwater can affect farmers' income negatively because it raises the cost of irrigation. An estimation based on data in Shanxi province of China finds that a tenfold increase in the price of water could reduce social welfare by 39 per cent (Fang and Nuppenau 2004). As farmers are the poorest people in Beijing, it seems unreasonable to increase the cost of irrigation (Ahmad 2000; Webber et al. 2008; Yang et al. 2003). From a policy perspective, the question is whether a groundwater charge can lead to an increase in rainwater consumption while maintaining farmers' income.

The objective of this chapter is to study how to increase the consumption of rainwater by charging for groundwater while not discouraging farming. This chapter consists of two parts: theoretical analysis and linear programming. In the theoretical analysis, the relation between the cost for groundwater and the consumption of rainwater is studied graphically by analysing the elasticity of groundwater. Many studies concerned with the price elasticity of irrigation water have found that when the price of water is lower than a threshold, irrigation water is price inelastic (Berbel and Gomez-Limon 2000; Salman and Al-Karablieh 2004; Scheierling et al. 2004; Yang et al. 2003). The cost for groundwater should be higher than the elasticity threshold so that the consumption of groundwater is sensitive to its cost. Hence increasing the cost of groundwater can raise the consumption of rainwater. If the elasticity threshold decreases, the cost of groundwater enabling the increase in rainwater consumption becomes lower. A low cost of groundwater would not discourage farming. Through linear programming, we try to find out how to decrease the elasticity threshold of groundwater, and then to determine a realistic charge for groundwater, which affects not only the consumption of rainwater positively but also the income of the farmers as little as possible.

5.2 A Theoretical Analysis using Graphs

It has been indicated in many studies that there are three thresholds for prices leading to different elasticity in groundwater demand change (Scheierling et al. 2004; Varela-Ortega et al. 1998; Yang et al. 2003). The groundwater can become inelastic at a too low price or at a too high price level. Only in the middle price range, the groundwater price is elastic. Figure 5.1 illustrates the relationship between the cost of groundwater and its consumption. It shows that when the cost is below a threshold value P_1, the consumption of groundwater is completely inelastic. Farmers respond very little or not at all to an increase in groundwater cost, but maintain the existing groundwater demand. The cost of groundwater is too low to induce farmers to achieve water saving or to use other sources of water such as rainwater for agricultural irrigation. When the cost is higher than P_1, water consumption changes with different water costs (Figure 5.1). Because of high elasticity, a small increase in the water cost can bring a large decrease in groundwater consumption. At this stage, farmers would use other water sources to supplement irrigation water. The elasticity decline when water cost is higher than P_2 (Figure 5.1). Given this situation, the consumption of groundwater is limited and farmers may decrease total water consumption through a series of water saving activities. They may change from water intensive crops to non-irrigated crops or crops requiring very little water.

Figure 5.1
The change in elasticity of groundwater in terms of different cost

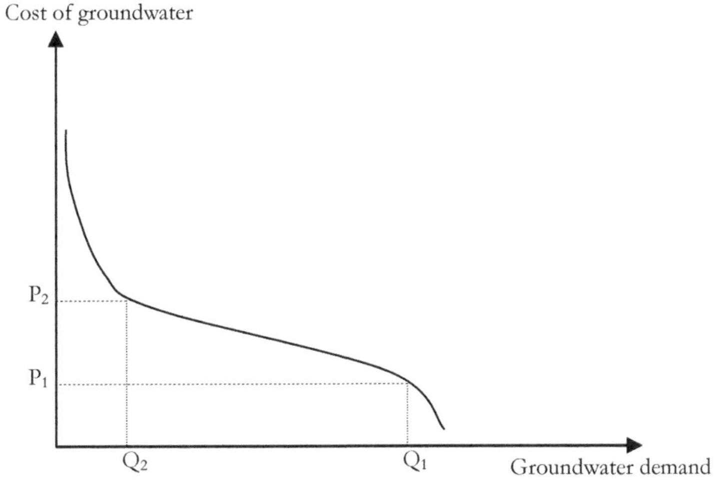

It is assumed that there are only two water resources for agricultural irrigation: groundwater and rainwater, and the water demand for irrigation is fixed. Rainwater is a substitute for groundwater, which means the changes in consumption of groundwater and rainwater are opposite. When groundwater consumption increases, rainwater consumption will decrease, and *vice versa*. Figure 5.1 describes the relationship between groundwater cost and consumption. When groundwater cost increases, the consumption of groundwater will decrease and meanwhile the consumption of rainwater will increase. Accordingly, based on Figure 5.1, the relationship between the cost of groundwater and the consumption of rainwater can be estimated and shown in Figure 5.2.

Figure 5.2
Relationship between groundwater cost and rainwater demand

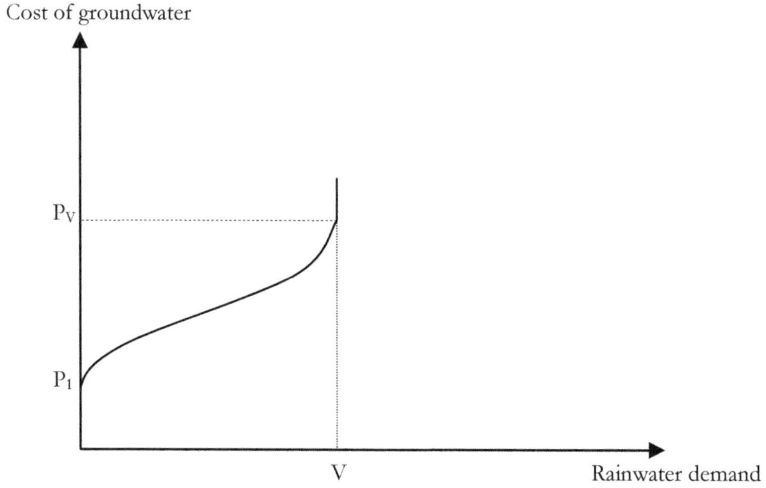

Figure 5.2 illustrates that when the cost of groundwater is lower than P_1, the consumption of rainwater tends to be zero. Farmers do not respond to the increased cost of groundwater and continue to use groundwater for agricultural irrigation. When the cost of groundwater is higher than the elasticity threshold P_1, the consumption of rainwater can rise with an increase in the cost of groundwater. Farmers respond to increased groundwater cost by reducing their consumption of groundwater. To keep the same amount of water for irrigation, farmers would raise rainwater consumption to supplement water supply. The existing crop distribution and water consumption remain the same. When the cost of groundwater increases to, or over $P_{v,}$ rainwater consumption reaches its potential maximum harvesting amount V. At this stage, rainwater harvesting systems can be fully utilized, while little groundwater is extracted for agricultural irrigation due to its high cost. Notice that the threshold value P_1 is the crucial point to improve the consumption of rainwater. As long as the cost of groundwater is higher than the threshold value P_1, the consumption of rainwater becomes sensitive to changes in groundwater cost. This implies the precondition for increasing the consumption of

rainwater is that the cost of groundwater is higher than the threshold value P_1.

Figure 5.3
Movement of rainwater demand lines

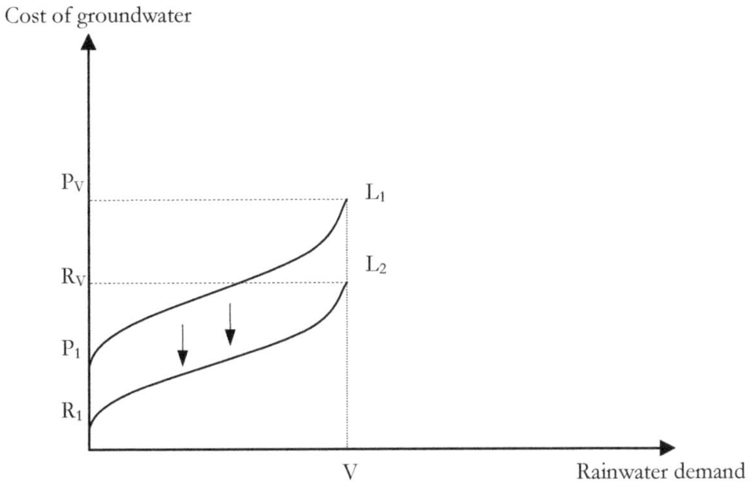

If the characteristics of using groundwater or rainwater are changed, the line in Figure 5.2 could move. Figure 5.3 illustrates the situation of the line of rainwater demands moving down. If the curve line of rainwater demands moves from line 1 (L_1) to line 2 (L_2), the threshold value decreases from P_1 to R_1. In the case of line 2 (L_2), when the groundwater cost is above R_1, the consumption of rainwater becomes sensitive to the change in the cost of groundwater. To reach the same rainwater consumption, the cost of groundwater in the case of line 2 (L_2) can be lower than the cost of groundwater in the case of line 1 (L_1). For example, to reach the potential maximum rainwater harvesting amount V, the cost of groundwater in line 1 (L_1) should be P_v while the cost of groundwater in line 2 (L_2) is R_v, which is smaller than P_v. As is mentioned previously, the condition of improving the consumption of rainwater is the cost of groundwater being higher than the elasticity threshold value. If the elasticity threshold value decreases, the cost of groundwater does not need to be so high to increase the consumption of rainwater. Low cost groundwater means the groundwater could be charged at a lower price and farmers' income is not affected negatively. So the decline in the elasticity threshold value helps to improve the consumption of rainwater while charging for groundwater at a low level.

According to the theory of substitution, the curve moving down means the cross elasticity increases. The responsiveness of the consumption of rainwater to the change of the cost of groundwater becomes higher. This can be taken in two ways: the change in characteristics of using groundwater, and the change in characteristics of rainwater harvesting. In this study, the characteristics of using groundwater do not change much since the capacity of wells and pumping cost tend to be similar. Hence the movement of curves represents the change in characteristics of rainwater harvesting systems, such as the size of the rainwater harvesting system, or subsidies for initial investments.

The graphic analysis demonstrates the relationship between the cost of groundwater and the consumption of rainwater. Rainwater consumption tends to be zero when the cost of groundwater is lower than the elasticity threshold. Once the cost of groundwater is greater than the elasticity threshold, farmers would increase the use of rainwater and decrease the consumption of groundwater. The threshold value is an important point to improve the consumption of rainwater. If the threshold value is low, smaller groundwater cost is required to enable the increase in rainwater use. We also find that the threshold cost could decrease through a movement of the curve line. The movement of the curve line for rainwater can be achieved by changing the characteristics of the rainwater harvesting system.

5.3 Linear Programming

We use linear programming to discover which characteristics of rainwater harvesting systems could lead to a change in the elasticity threshold of groundwater.

5.3.1 Data

The data source for the analysis of this chapter is similar to the data source of Chapter 4, and is based on interviews with managers of rainwater harvesting projects in the rural areas of Beijing. Additionally, some relevant data are taken from the literature (Tian et al. 2007; Wang et al. 2007; Yang and Abbaspour 2007).

5.3.2 Linear programming model

From the point of view of farmers, the objective of linear programming is to obtain the maximum net profit. We assume that increasing water consumption in agricultural irrigation could raise the production of crops. In this case, there are two alternative irrigation water sources: groundwater and rainwater. The objective function (Equation 5.1) illustrates that the net profit of production is determined by the consumptions of groundwater (q_1) and rainwater (q_2).

Objective function:

$$\text{Maximum } z = b(q_1 + q_2) - c_1 q_1 - c_2 q_2 \tag{5.1}$$

where, z is the net benefit of the project; b is the unit profit of crops (Yuan/m^3). The unique profit of crop income (b) in Beijing could be set at 6 Yuan/m^3, using data by Wang et al. (2007). The main crops cultivated in the rainwater reuse systems studied are cucumber, tomato, lettuce and marrow; q_1 is the consumption of groundwater (m^3); q_2 is the consumption of rainwater (m^3); c_1 is the unit cost of groundwater (Yuan/m^3); c_2 is the unit cost of rainwater (Yuan/m^3).

Constraints functions:

$$o \leq q_1 \leq Q \tag{5.2}$$

$$q_1 + q_2 \leq Q \tag{5.3}$$

$$o \leq q_2 \leq V \tag{5.4}$$

Equations 5.2, 5.3 and 5.4 illustrate the constraints on the consumption of groundwater (q_1) and rainwater (q_2). Despite the fact that there is no limit on the amount of groundwater pumped up from a well, its consumption (q_1) cannot be more than the quantity required for irrigation (Q) (Equation 5.2). Similarly, the total consumption from the two water sources also cannot be greater than the quantity required for irrigation (Q) (Equation 5.3). Moreover, the collected rainwater amount (q_2) is restricted by the potential maximum collected rainwater amount (V) (Equation 5.4).

The quantity required for irrigation (Q) relies on the water quantity required for unit irrigating area (L) and the area under irrigation (S), shown in Equation 5.5. According to Wang et al. (2007), L is 0.3 m^3/m^2, S is 1.17 million m^3. So the volume of the potential required irrigating water (Q) in the studied rainwater harvesting systems can be calculated to be 350,000 m^3.

$$Q = L \times S = 0.3 \times 1.17 \times 10^6 = 0.35 \times 10^6 m^3 \tag{5.5}$$

The potential maximum collected rainwater amount (V) depends on the volume of rain falling (k_p), the area where rainwater is collected (A), and the coefficient of effective rainwater harvesting (K), shown in Equation 5.6. According to two studies, k_p equals 452 mm, A is 9.3 million m^2, K is 0.8 (Tian et al. 2007; Wang et al. 2007). Hence the maximum collected rainwater volume (V) is calculated to be 340,000 m^3 (Equation 5.6).

$$V = k_p \times A \times K / 1000 = 452 \times 9.3 \times 10^6 \times 0.8 \div 1000 = 0.34 \times 10^6 m^3 \quad (5.6)$$

5.3.3 Graphical solution of linear programming

In terms of the constraints functions, we can determine the region of feasible solutions. As $q_1 + q_2 \leq Q$, the possible values of q_1 and q_2 lie below the line of Q_1Q_2. But q_2 is smaller than V. The determination of Equations 5.5 and 5.6 illustrates that the V value is smaller than the Q value. So the feasible values of q_1 and q_2 are limited to the dark area of Figure 5.4. The optimal solution will be a point on the frontier of the region of all feasible solutions.

Figure 5.4
Region of feasible solutions

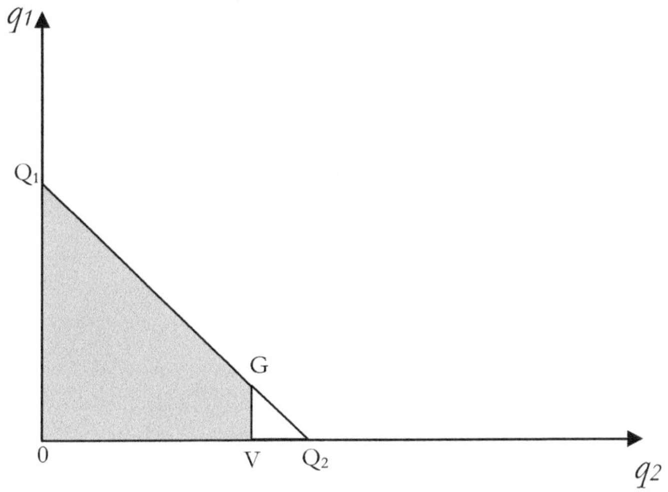

The objective function can be presented graphically using isoprofit lines. In this case, the isoprofit line is denoted by line *AB*, shown in Figures 5.5, 5.6 and 5.7. The optimal solution can be found by the point of tangency of the frontier of the region of feasible solutions to the highest possible isoprofit curve (Koutsoyiannis 1979). So the optimal solution depends on the slope of the isoprofit lines.

In this study, the objective function is $z = b(q_1 + q_2) - c_1q_1 - c_2q_2$, which can be changed to be $q_1 = \frac{z}{b-c_1} - \frac{b-c_2}{b-c_1} \times q_2$. Thus the slope of the isoprofit line is $\frac{b-c_2}{b-c_1}$. The possible frontier of the region of feasible solutions includes two points: Q_1 and G and one line: Q_1G. As the line of Q_1G is the boundary line of the constraint: $q_1 + q_2 \leq Q$, its slope equals 1. Following the change of the slope of the isoprofit lines, three kinds of optimal solutions could appear.

Figure 5.5
Optimal solution 1 (the isoprofit slope < 1)

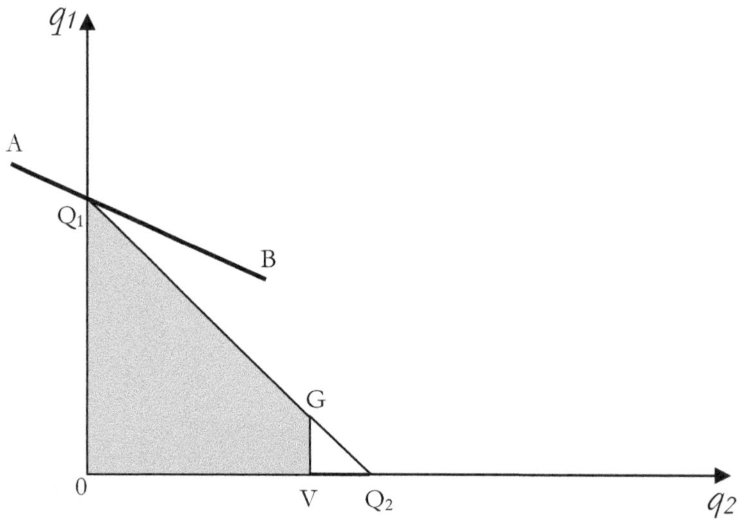

First, if the slope of the isoprofit lines is smaller than 1, the point of tangency is the point of Q_1, shown in Figure 5.5. At the point of Q_1, $q_1 = Q$

and q_2=0, which means that the consumption of rainwater is zero while the consumption of groundwater reaches its maximum. In this situation, $\frac{b-c_2}{b-c_1}<1$, namely $c_2>c_1$. So when the unit cost of groundwater is smaller than the unit cost of rainwater, the consumption of rainwater is zero.

Second, if the slope of the isoprofit lines equals 1, all the points in the line of Q_1G could be the optimal solutions, shown in Figure 5.6. Given this situation, the consumption of rainwater could range from 0 to V. The slope of the isoprofit lines equals 1, which means $\frac{b-c_2}{b-c_1}=1$, namely $c_1=c_2$. So when the cost of groundwater and the cost of rainwater are the same, farmers may start to use rainwater.

Figure 5.6
Optimal solution 2 (the isoprofit slope = 1)

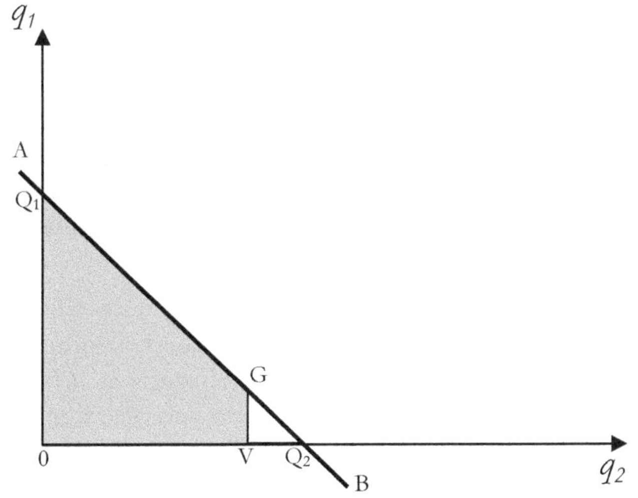

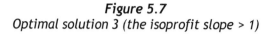

Figure 5.7
Optimal solution 3 (the isoprofit slope > 1)

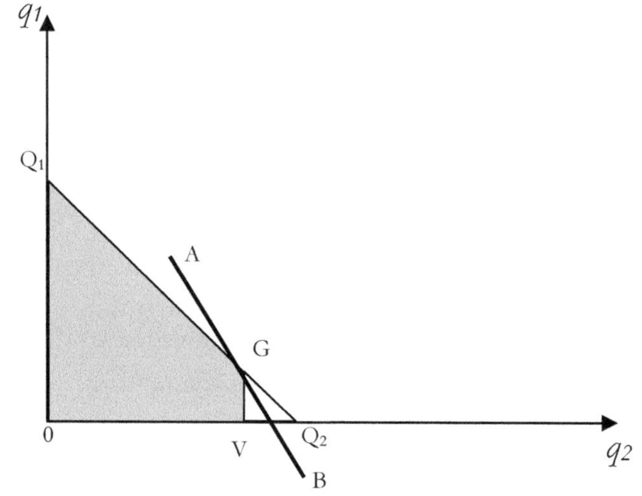

Third, if the slope of the isoprofit lines is greater than 1, the point of tangency is the point of *G* shown in Figure 5.7. At the point of *G*, the consumption of rainwater is *V* and the consumption of groundwater is (*Q*-*V*). In this situation, $\frac{b-c_2}{b-c_1} > 1$, namely $c_2 < c_1$. So when the unit cost of groundwater is greater than that of rainwater, farmers would decrease the consumption of groundwater and increase the consumption of rainwater.

Thus different values of the cost of groundwater and rainwater will lead to different optimal solutions for the consumption of both water sources. Since zero consumption of rainwater is not suitable to the objective of increasing rainwater consumption, the first optimal solution is not considered in this study. The second optimal solution shows that when the cost of groundwater equals the cost of rainwater, the consumption of rainwater could increase. When the cost of groundwater is greater than the cost of rainwater, rainwater consumption reaches the maximum potential value, shown in the third optimal solution. Hence the cost of rainwater could be regarded as the elasticity threshold of groundwater.

Equation 5.7 shows that the cost of rainwater includes the initial investments (c_I), and operation and maintenance cost (c_o), which are affected by two major factors: the size of rainwater harvesting systems and subsidies. In terms of the economies of scale, the large systems have a small unit initial capital input, while the small systems have a large unit capital input. The unit capital input means the total capital input divided by the capacity of the rainwater harvesting plant. In the same way, the unit operation and maintenance cost of the large systems is lower than that of the small systems. Additionally subsidies could effectively decrease the initial expenditure paid by farmers to reduce the cost of using rainwater harvesting systems. The calculation of the unit initial cost of rainwater harvesting is shown in Equation 5.8

$$c_2 = c_I + c_o \tag{5.7}$$

$$c_I = u_I\left(\frac{1-r_1}{L} + \frac{r_2}{L}\right) \tag{5.8}$$

where, u_I is unit capital investment; r_1 is residential ratio of rainwater harvesting system; r_2 is replacement ratio of rainwater harvesting system; L is lifetime of system. According to the literature, r_1 is 4%, r_2 is 0.25 per cent and L is 20 years (Yang and Abbaspour 2007).

As mentioned the cost of rainwater could be regarded as the elasticity threshold of groundwater. So the decrease in the cost of rainwater means a decline in the elasticity threshold of groundwater. The cost of rainwater is mainly affected by the size of rainwater systems and subsidies. Hence the elasticity threshold of groundwater could be affected by the size of rainwater harvesting systems and subsidies indirectly. It confirms the argument of the precious section that the elasticity threshold of groundwater could be affected by the characteristics of rainwater harvesting.

Equation 5.9 shows that the cost of using groundwater consists of two parts: pumping cost (c_e) and a water charge (p_u). The pumping cost (c_e) depends on the depth of a well, which is generally 80 metres. According to the interviews with project managers, the average unit energy cost of pumping groundwater (c_e) is calculated to be 0.13 Yuan/m^3. As the pumping cost is a fixed value, the change in the cost of using groundwater depends on the groundwater charge.

$$c_1 = c_e + p_u \tag{5.9}$$

5.3.4 Optimal threshold and groundwater charge

Two major characteristics: the size of rainwater harvesting systems and the subsidies relate to the cost of rainwater harvesting. According to interviews, there are three sizes of rainwater harvesting systems: 50 m^3, 450 m^3 and 1,300 m^3 and two kinds of subsidies: 25 per cent and 75 per cent. For comparative analysis, data is put into the equations. The comparative analysis is carried out in terms of different sizes of rainwater harvesting systems and different kinds of subsidies.

Table 5.1
The data of different system sizes without subsidies (Yuan/m^3)

Size	c_o	u_l	c_1
50 m^3	2.41	140	6.7
450 m^3	0.51	92	4.4
1300 m^3	0.41	77	3.7

Tables 5.1 and 5.2 indicate the data of initial investments and operation and maintenance cost of three system sizes, obtained from interviews with project managers. Normally subsidies only are provided to initial investments and no subsidies to operation and maintenance cost. Table 5.2 shows that the operation and maintenance cost (c_o) of each size system is the same and the capital investment (u_1) decreases, given the situation of providing subsidies.

The previous analysis indicates that the cost for rainwater could be regarded as the value of the elasticity threshold of groundwater. After putting the data from Tables 5.1 and 5.2 into Equation 5.7, the cost of the threshold point can be obtained (Table 5.3).

Table 5.2
The data of different system sizes with subsidies (Yuan/m³)

Sizes and Subsidies	c_o	u_l	c_1
50 m³			
25 % of investments	2.41	105	5
75 % of investments	2.41	35	1.7
450 m³			
25 % of investments	0.51	69	3.3
75 % of investments	0.51	23	1.1
1300 m³			
25 % of investments	0.41	58	2.8
75 % of investments	0.41	19	0.9

The results of the threshold values of groundwater according to different sizes and subsidies of rainwater harvesting systems are listed in Table 5.3. We find that the threshold values of groundwater are decreasing with the increase in the size of rainwater harvesting. For example, given the situation of no subsidies, the threshold value is 9.11 Yuan/m^3 if the size of rainwater systems is 50m^3; while the threshold decreases to 4.91 Yuan/m^3 if the size is enlarged to 450m^3; then the threshold reduces to 4.11 Yuan/m^3 if the size is 1,300 m^3. Moreover, an increase in the subsidies on initial investments of rainwater harvesting could decrease the value of threshold of groundwater. For example, provided the size is 450 m^3, the threshold value could decrease from 3.81 to 1.61 Yuan/m^3 when the subsidies increase from 25 to 75 per cent (Table 5.3). Therefore we can draw the conclusion that the threshold of groundwater can decrease if the size of rainwater harvesting systems is greater or the subsidies on initial investments increase.

Table 5.3
Threshold of groundwater in terms of different system sizes and subsidies

Sizes and Subsidies	Threshold of groundwater (Yuan/m^3)
50 m^3	
0 subsidies	9.11
25 % of investments	7.41
75 % of investments	4.11
450 m^3	
0 subsidies	4.91
25 % of investments	3.81
75 % of investments	1.61
1300 m^3	
0 subsidies	4.11
25 % of investments	3.21
75 % of investments	1.31

Linear programming shows that if the cost of groundwater is greater than the cost of rainwater, namely $c_1 > c_2$, this would lead to the maximum consumption of rainwater. If the data in Tables 5.1 and 5.2 are put into Equations 5.7 and 5.9 and the function of $c_1 > c_2$, the realistic groundwater charges in terms of various sizes of rainwater harvesting systems and different kinds of subsidies could be obtained (Table 5.4).

Table 5.4 shows that the realistic groundwater charge decreases when the size of rainwater harvesting systems becomes greater. For example, given the situation of no subsidies, the realistic groundwater charge should be greater than 8.98 Yuan/m^3 if the size of rainwater system is 50 m^3 while the realistic groundwater charge should be greater than 3.98 Yuan/m^3 if the size of rainwater system is 1300 m^3. Table 5.4 also shows that if the subsidies for rainwater harvesting systems increase, the groundwater charge could be less. For a rainwater system of 450m^3, 25 per cent of initial investment is subsidized, the realistic groundwater charge should be greater than 3.68 Yuan/m^3 in order to increase consumption of rainwater. But if the subsidies increase to 75 per cent, at $450m^3$, the groundwater charge is greater than 1.48 Yuan/m^3 (Table 5.4). This implies that when the rainwater harvesting system is greater and the subsidies are higher, the realistic groundwater use could be smaller. For example, a large rainwater system — 1,300 m^3 and 75 per cent of initial investments subsidized — the realistic groundwater charge should be greater than 1.18 Yuan/m^3 (Table 5.4).

Table 5.4
Realistic groundwater charge for different system sizes and subsidies

Subsidies	**Realistic groundwater charge**
50 m^3	
0 subsidies	> 8.98 Yuan/m^3
25 % of investments	> 7.28 Yuan/m^3
75 % of investments	> 3.98 Yuan/m^3
450 m^3	
0 subsidies	> 4.78 Yuan/m^3
25 % of investments	> 3.68 Yuan/m^3
75 % of investments	> 1.48 Yuan/m^3
1300 m^3	
0 subsidies	> 3.98 Yuan/m^3
25 % of investments	> 3.08 Yuan/m^3
75 % of investments	> 1.18 Yuan/m^3

5.4 Conclusions

This chapter uses theoretical analysis and linear programming to study how to increase the consumption of rainwater by charging for groundwater while not discouraging farming. The theoretical analysis provides the elasticity threshold for a groundwater charge. If the groundwater cost were lower than the threshold, farmers would have few incentives to change their source from groundwater to rainwater. If the groundwater cost were higher than the threshold, the consumption of rainwater would increase. The theoretical analysis also shows that the elasticity threshold of groundwater could move down following the change in the characteristics of rainwater harvesting systems. The linear programming illustrates that increasing subsidies and enlarging the size of rainwater harvesting systems can decrease the elasticity threshold of groundwater. Decreasing elasticity threshold of groundwater implies that the cost of groundwater enabling the increase of rainwater consumption becomes lower. Hence a realistic groundwater charge has been determined by taking into account different sizes of rainwater harvesting systems and the subsidies on the initial investments for the rainwater systems.

The price mechanism alone is not enough to increase rainwater consumption. Increasing water price may raise farmers' incentives to use rainwater, but it also could lead to a significant decrease in farmers' income. It is important to reach a balance of effectively increasing rainwater demand while keeping an eye on farmer's benefits. The results of this chapter indicate that increasing subsidies for initial investments and enlarging the size of rainwater harvesting systems can both help to achieve this delicate balance.

Notes

[4] Chapter 5 is submitted to the journal, *Water*, 2010, under review.

6 Decisive Factors Affecting the Operation of Rainwater harvesting Plants[5]

6.1 Introduction

Chapters 4 and 5 found that enlarging size and increasing subsidies could help to improve the performance of rainwater harvesting. However, according to interviews with plant managers, operation of many rainwater harvesting plants was unsuccessful despite the scale and subsidies. Hence the size of systems and subsidies may not be critical factors affecting the operation of the rainwater harvesting systems. In Chapter 6, the decisive factors affecting the operation of the rainwater harvesting plant will be analysed.

Only 20 per cent of the constructed rainwater harvesting systems in Beijing operate continually and most of the remaining systems operate intermittently, while some systems are in ruins (source: interviews with the officials of the Beijing Agro-Technical Extension Center). This is an 80 per cent failure rate for rainwater harvesting plants in rural areas of Beijing. Rainwater harvesting plants in the rural areas of Beijing obtain most of their technical support from the Beijing Agro-Technical Extension Center, a professional organization focusing on providing agricultural technologies to farmers. However the technological problems may not be the major reasons for the operational failure of some rainwater harvesting plants in Beijing.

Researchers have conducted numerous studies on developing rainwater harvesting for agricultural irrigation, especially in Sub-Saharan Africa, the Middle East and Asia (Hatibu et al. 2006; Li et al. 2000; Mushtaq et al. 2007; Pandey 1991; Tian et al. 2002). In the existing research, the technical improvement of rainwater harvesting is highlighted because the water yield from early rainwater harvesting techniques is insufficient for modern agricultural purposes (Hatibu et al. 2006; Oweis and Hachum 2006). Technical improvement does effectively facilitate the operation of a new system, but does not determine its successful operation. Whether the operation of a system is a success or failure depends on various factors, technological and non-technological. For instance, inefficient management, low cost recovery or negative social impact all could lead to the failure of the operation of a system. The non-technological factors affecting the operation of rainwater harvesting in the literature are rarely discussed (He et al. 2007).

This chapter aims to find the decisive factors, which determine the operational success or failure of a rainwater harvesting plant in Beijing. Based on literature and interviews with plant managers, ten possible factors explaining failure were chosen. It is assumed that the operational status of a rainwater harvesting system in Beijing will be affected by these ten factors including technological and non-technological. The methodology of rough set analysis is used to identify the decisive factors determining the operational success or failure of rainwater harvesting systems in Beijing. Rough set analysis is a technique of artificial intelligence (Pawlak 1982; Slowinski 1991).

Ten representative plants with different features and operational status are chosen for the research. The details are presented in Section 6.2. Ten factors are introduced and explained in Section 6.3. The rough set method and how to derive the decisive factors are presented in Section 6.4. Finally, discussion of the results and the conclusions are in Section 6.5 and 6.6 respectively.

6.2 A Concise Overview of the Cases

As mentioned in Chapter 4, we did extensive fieldwork on rainwater harvesting in the rural areas of Beijing. The fieldwork was undertaken in close collaboration with the Chinese Academy of Sciences and the Beijing Agro-Technical Extension Center. Although there are around 600 rainwater harvesting plants in the rural areas of Beijing, parts of these plants are supervised and subsidized by the Beijing Agro-Technical Extension Center and others are supervised by other organizations such as the Beijing Water Saving Office. The number of rainwater harvesting plants supervised by the Beijing Agro-Technical Extension Center is around 300. Because of the cooperation, fieldwork focuses on rainwater harvesting plants supervised by the Beijing Agro-Technical Extension Center.

Due to time limitations and Beijing being a large city, only 30 rainwater harvesting plants were considered. Officials of the Beijing Agro-Technical Extension Center suggested these 30 plants. They include plants with good and poor operations. The recommendations from the Beijing Agro-Technical Extension Center depended on whether it had data on the plants.

Although interviews were conducted in 30 plants, only ten plants were chosen for the analysis. The criteria of selection are construction completed for at least two years and sufficient data on the plant is available. A concise overview of these cases is as follows.

P1: A farming household plant

This plant is located in the Changping district of Beijing city, owned by a private farming household. The capacity of the plant is 60 m^3. It was constructed in 2007, but it has been used for only one year. The farmers of this plant consider the rainwater quality unsuitable for irrigation, and think using groundwater is more convenient.

P2: Jin Liu Huan plant

This plant is located in the Changping district, managed by a state-owned company. The capacity of the plant is 1,000 m^3. This plant was built in 2006. Unlike other plants constructed underground, the plant is in a room. Although it has no technological problems, the plant is used occasionally. The purpose of this plant is for demonstration, not for practical use. Currently, it is still convenient to access groundwater.

P3: Xiao Tang Shan plant

This is a privately-owned plant, located in the Changping district. It has a larger capacity, 2,000 m^3. It was built in 2005, earlier than the other plants. The storage tank is at the surface without any cover, which causes phytoplankton growth in the tank. The phytoplankton becomes stuck in the pipes and the pump leading to plant interruptions.

P4: Chang Shang Cun plant

The plant is located in the Shunyi district and belongs to a household. The capacity of the plant is 500 m^3. It was constructed in 2007, and has been only used half of the year. The manager considers that the water quality is not good enough for irrigation. There is a serious technological problem in this plant, but the plant manager does not want to solve it since it requires additional expenditures to improve the system.

P5: The Agro-Technical Extension Center of Shunyi district plant

This plant was constructed by the Agro-Technical Extension Center of Shunyi district. There are subsidies for the operation and maintenance cost. It was built in 2006. The capacity of the plant is 250 m^3. It is a state-owned plant. The manager thinks that the rainwater quality is low, and the technology needs to be improved. The attitude of the plant manager implies that the plant could not operate in a sustainable way.

P6: Tao Li Cun plant

The plant is located in the Miyun district, a mountainous area. It is very difficult to obtain groundwater in this district. Farmers changed to a new well after the existing well no longer produced water. In the summer, because limited groundwater could be obtained from the personal well, farmers get groundwater from a collective big well based on quota. This rainwater harvesting plant is the only one operating successfully. It was built in 2006, subsidized largely by the government. The capacity of this plant is 60 m^3, which effectively alleviates the pressure of irrigation supply of a household farm area (around 933 m^2).

P7: Li Ge Zhuang plant

This plant is located in the Miyun district, owned by a household. The plant was constructed in 2007. The capacity of the plant is 50 m^3. In 2008, a new well was drilled in the village of this plant and farmers can get water from the new well without any charge. So normally the irrigation water of the plant is taken from the new well. But in periods of drought, the water available in the well decreases so that there is insufficient water for the village. In this situation, the plant manager would use the rainwater harvesting system to obtain the required water. Hence the Li Ge Zhuang plant is only used in periods of drought.

P8: Hong Ke Farm plant

The plant is located in a state-owned farm in the Fangshan district where it is easy to access the groundwater resource. The plant was built in 2006, and the capacity is 750 m^3. In this plant, the storage tank is well designed and located in a room. But it is used for half the year and then stops operation. Although the manager found serious technological problems they went unrepaired for two years.

P9: Shi Lou Cun plant

The plant is located in the Fangshan district, and is owned by a farm household. It was constructed in 2008, and the capacity is 75 m^3. As the storage tank is constructed at a high level, the rainwater cannot go into the storage tank effectively. In 2009, the rainwater harvesting plant manager was changed because the farm was rented to another farmer. Due to a technological problem and the farmer's doubts about rainwater quality, the new plant manager stopped operation of this plant.

P10: The Agro-Technical Extension Center of Daxing district plant

The plant is located in the Daxing district, managed by the Agro-Technical Extension Center of Daxing district. The plant was constructed in 2008 and the capacity is 200 m^3. Although the plant was built recently, the storage tank has serious leakage problems. The collected water is insufficient for irrigation requirements. Hence the plant is only used half the year and then operation stops.

6.3 Chosen Factors

Various factors might affect the operation of rainwater harvesting systems. Based on the literature and interviews with plant managers, ten factors are chosen for this study: 1) ownership, 2) farmers perception of rainwater, 3) doubts about the rainwater quality, 4) location, 5) availability of groundwater, 6) size of storage, 7) irrigation method, 8) technical problems, 9) subsidies for initial investment, and 10) subsidies for operation and maintenance. These factors are responsible for the success or failure of rainwater harvesting plants in Beijing. The details for each factor appear below.

6.3.1 Ownership

Who is responsible for ensuring the daily operation and performance of management functions is very important to the operation of a water system (Mushtaq et al. 2007). Different types of ownership leads to different incentives to manage and operate the systems.

In this study, some rainwater harvesting plants are managed privately and governmental organizations as well as state-owned farms manage others. Generally the incentive of private plant managers to operate a rainwater harvesting system comes from the decrease in irrigation cost or the improvement of income. But the incentives of the state-owned plant managers do not depend on cost or income, which may lead to continuing operation of state-owned plants even though they are losing money.

6.3.2 Farmers' perception of rainwater

Compared with using groundwater, using rainwater for agricultural irrigation is a 'new system' to the local farmers. Using rainwater is usually considered more complex than using groundwater although the technology of rainwater harvesting is simple.

The farmers' negative perception of rainwater harvesting could affect the successful operation of a rainwater harvesting system. Farmers' positive perceptions of rainwater are a significant factor promoting the success of rainwater harvesting plants (He et al. 2007; Song et al. 2009). Some farmers have a positive perception of rainwater as rainwater can supplement irrigation water supply. On the other hand, some farmers consider it is unnecessary to use rainwater due to various non-scientific reasons, such as that it is too troublesome to do things differently. We learned about the farmers' perception of rainwater during the interviews.

6.3.3 Doubts about rainwater quality

Many people doubt the quality of harvested rainwater (Song et al. 2009). Most of them point out that possible pollutants from the atmosphere involved in water could cause negative effects on crops. Beijing is a mega city with 1.3 million people and a large number of factories. Particles, heavy metals and organic air pollutants contaminate the atmosphere (Helmreich and Horn 2009).

Some farmers doubt if the rainwater quality in Beijing is suitable to irrigate farmlands. There is no systematic information proving whether rainwater in Beijing is suitable although some experiments have been implemented. The doubts concerning rainwater quality could result in a resistance against using rainwater for irrigation.

6.3.4 Location

Table 6.1
Depth of groundwater in northern and southern plants (2007-2008)

	P1	P2	P3	P4	P5	P6	P7	P8	P9	P10
Depth of groundwater	28	28	28	32	32	30	30	13	13	20

Plants P1-P7 lies in the north, P8-P10 lies in the south.
Source: Beijing Water Bulletin (2007-2008)

He et al. (2007) find that the location of a plant has significant impact on the operation of selected rainwater harvesting plants. In this study, ten rainwater harvesting plants were selected for further study, of which seven plants are located in the northern area of Beijing and three plants are located in the southern area of Beijing. The north of Beijing is mostly a mountainous area, and the south is a relatively flat area. Accordingly, the groundwater level in the north is lower than in the south. Table 6.1 indicates the depth of groundwater levels of northern plants compared to southern plants. It means farmers living in the northern area need to pump groundwater from deeper wells than farmers living in the south do.

Consequently higher cost is needed when pumping water from deeper wells. Farmers living in the north may pay more to access groundwater so that they could be more inclined to use rainwater.

6.3.5 Availability of groundwater

In this study, the availability of groundwater resources for various plants is different. Groundwater resources are extremely limited in some places. Groundwater only emerges in certain wells in the summer season and the wells remain dry in other periods. These places could be regarded as places of water scarcity. However, the groundwater resources could be sufficient in other places where it is easy to access groundwater through pumping from a well. This could be regarded as places of water sufficiency.

Many studies have pointed out that rainwater harvesting could effectively relieve the pressure of water scarcity (Helmreich and Horn 2009; Li et al. 2000; Oweis and Hachum 2006). The farmers in places of water scarcity would have more incentives to use new water resources than the farmers located where there is water sufficiency.

6.3.6 Size of storage tank

The size of rainwater harvesting systems determines the investment and the expenditure of operation and maintenance paid by farmers. In terms of economies of scale, the large-scale systems have a small initial capital input per unit, while the small-scale systems have a large capital input per unit (Goel and Kumar 2004). In the same way, the unit operation and maintenance cost of the larger systems is lower than the cost of the small systems. However, the total initial investment cost and operation and maintenance cost of the larger system is more than the cost of the small systems. In this study, the storage tanks of the rainwater harvesting systems have been classified into three sizes: small (50-100 m^3), medium (450-800 m^3) and large (1,300-2,000 m^3).

6.3.7 Irrigation methods

Irrigation methods normally include drop irrigation and drip irrigation. Drop irrigation as the traditional method is largely used by farmers in Beijing. However, drip irrigation is currently promoted in Beijing, since the drip irrigation method could effectively decrease water consumption. As high initial expenditures for facilities are needed for drip irrigation, not all farmers can afford it.

Irrigation methods affect operation of rainwater harvesting systems (Berthelot and Robertson 1990). In rainwater harvesting systems, using drip irrigation always leads to the problem of muddle stuck. The water distribution pipe used in drip irrigation is very narrow. The process of water collection is that rainwater flows into the storage tank with muddle and some natural rocks. When pumping water from the storage tank, the muddle and rocks flow with the water and are stuck in the narrow pipes thereby hampering the operation of the rainwater harvesting system.

6.3.8 Technical problems

In the literature, the most important issue concerning rainwater harvesting is the necessary technical improvements (Hatibu et al. 2006; Oweis and Hachum 2006). Usually the Beijing Agro-Technical Extension Center or the Beijing Water Saving Office designs the facilities and construction. As the rainwater harvesting systems in Beijing are at an early stage, various technical problems are emerging during operation. All the cases studied have problems related to the facilities and construction.

In terms of operation of the systems, the technical problems could be classified by three levels: 1) Repairable problems: some facilities are broken but they are repairable. For example, the pump is broken or a pipe is stuck. 2) Simple problems: the facilities should be changed or parts of the construction of the plant need changing. For instance a shallow ditch is changed to a deep ditch in order to improve rainwater collection. 3) Serious problems: the design of the construction of the plant is not suitable to operate continually. For example, in the Shi Lou Cun plant (P_9), the rainwater storage tank was built on top of a hill, which seriously decreases the volume of rainwater going into the storage tank.

6.3.9 Subsidies for initial investment

Subsidies for initial investment can effectively help to limit farmers' pressure on expenditures (He et al. 2007). For larger systems, more money is required as initial investment. For example, for the small size rainwater harvesting systems with a capacity of 50 m^3, the initial investment is around 27,000 Yuan including expenditures on construction and other facilities. The capacity of 50 m^3 is only suitable for irrigation water needs of a household. The total income of a small farming household in Beijing averages around 10,000 Yuan per year. The initial investment is almost three times the income of a household. If there are no subsidies, it is very difficult for farmers to afford such an initial investment.

Most rainwater harvesting systems are provided with subsidies of around 50 to 100 per cent of the initial investment although some systems are subsidized with less than 50 per cent of initial investment.

6.3.10 Subsidies for operation and maintenance

Many researchers study the operation and maintenance cost of rainwater harvesting (Berthelot and Robertson 1990; Mushtaq et al. 2007; Oweis and Hachum 2006). Whether the rainwater harvesting systems have sound maintenance for cost recovery influences sustainable operation of the systems. Insufficient input in maintenance could hamper the operation of rainwater harvesting systems. Subsidies for operation and maintenance benefit successful operation because total cost is lower.

In this study, the rainwater harvesting systems managed privately usually do not obtain subsidies for operation and maintenance; while systems managed by the state-owned farm or governmental organizations are more likely to get subsidies from the government.

6.4 Rough Set Analysis

Rough set, being part of set theory, is a mathematical method to synthesize the approximation of concepts from the data (Pawlak 1982; Slowinski 1991). The assumption of the rough set analysis is that every set could be roughly defined (Zhang and Xu 2009). For an extensive theoretical description of the rough set analysis, we refer to Pawlak (1982) and Slowinski (1991).

The rough set analysis is rarely applied in water resources (Barbagallo et al. 2006; Zhang and Xu 2009), although the rough set analysis has been extensively applied in other scientific fields, such as medicine, economics, environmental studies, software engineering and urban studies (Baycan-Levent and Nijkamp 2009; Nijkamp et al. 2000, 2002; Wu et al. 2004).

The objective of the study is to find the decisive factors determining the operational success or failure of rainwater harvesting systems in Beijing. Given the small sample (ten cases) and that the information collected is partly qualitative and partly quantitative data, the rough set analysis is a suitable method to derive the decisive factors through information classification and data mining.

In the social science research, most information on the selected plants is qualitative in nature, and some information is a mixture of qualitative and quantitative data. Similarly, the ten chosen factors explained in Section 6.3 could not be presented quantitatively in its entirety. It is necessary to synthesize this information for further information classification and data mining. First, the relevant information on ten factors is codified with 1, 2 or 3, to establish a consistent database, which is shown in Table 6.2. Second, because there are ten rainwater harvesting plants chosen for this study, a data matrix can be formed based on ten factors on ten plants (shown in Table 6.3). We find in Table 6.3 that the coded values of ten factors of ten plants are different and each factor is denoted by A_i (i=1,2,3...10) and the status of operation is denoted by D.

Table 6.2
Description and coded values of the factors

Factors	Description and Coded Value
A_1: Ownership	1: Private
	2: State-owned
A_2: Farmers' perception of rainwater	1: Negative
	2: Positive
A_3: Doubts about rainwater quality	1: Yes
	2: No
A_4: Location	1: North
	2: South
A_5: Availability of groundwater	1: Sufficient
	2: Scarce
A_6: Size of storage	1: Small
	2: Medium
	3: Large
A_7: Irrigation methods	1: Drop irrigation
	2: Drip irrigation
A_8: Technical problems	1: Reparable problems
	2: Simple problems
	3: Serious problems
A_9: Subsidies for initial investment	1: 51-100% of initial investment;
	2: 0-50% of initial investment.
A_{10}: Subsidies for operation and maintenance	1: Yes
	2: No
D: Status of each plant	1: Stopped
	2: Interrupted
	3: Continuous

The last line of Tables 6.2 and 6.3 indicates the coded values of the status of the operation of each plant. In terms of interviews, the status is stopped, interrupted or continuous. The status represents the situations of constructed rainwater harvesting systems in Beijing. The status of 'stopped' means there are no plant operations or the plant is in ruin, which is presented by '1'. The status of 'interrupted' means the plant operates or stops irregularly, which is presented by '2'. The status of 'continuous' means the plant operates successfully and continually, which is presented by '3'.

Table 6.3
Codified data matrix of ten factors and the operational status of ten plants

	P_1	P_2	P_3	P_4	P_5	P_6	P_7	P_8	P_9	P_{10}
A_1: Ownership	1	2	1	1	1	1	1	2	1	2
A_2: Public perception of alternative water resources	1	2	2	2	1	2	2	1	1	1
A_3: Doubts about rainwater quality	1	2	2	1	2	2	2	1	1	1
A_4: Location	1	1	1	1	1	1	1	2	2	2
A_5: Availability of groundwater sources	1	1	1	1	1	2	1	1	1	1
A_6: Size of storage	1	3	3	2	2	1	1	2	1	2
A_7: Irrigation methods	2	2	1	1	2	1	1	2	2	1
A_8: Technical problems	2	1	2	2	3	1	2	2	3	3
A_9: Subsidies to initial investment	1	1	1	1	1	1	1	2	1	2
A_{10}: Subsidies to maintenances expense	2	1	1	2	1	2	2	1	2	2
D: Status of each plant	1	2	2	1	2	3	2	1	1	1

* Pi (i=1,2,3...10) means the plant i

Since the operational status of each plant is assumed to be affected by ten factors, these ten factors (A_1, A_2, A_3 ... A_{10}) could be regarded as condition attributes and the status of each plant (D) could be regarded as decision attributes. We aim to find the causal links between condition attributes and decision attributes to identify the decisive factors in the operation of rainwater harvesting plants.

It is assumed that these ten plants P_1, P_2, P_3 ... P_{10} belong to a set U, namely

$$U = \{P_1, P_2, P_3, P_4, P_5, P_6, P_7, P_8, P_9, P_{10}\}$$

Table 6.2 shows that each condition attribute has its coded value. For example, A_1 (ownership) could be '1' (private owned) or '2' (state owned), or A_6 (size of storage) could be '1' (small) or '2' (medium) or '3' (large). So there are ten sets listed as follows.

$A_1 = \{1, 2\}$

$A_2 = \{1, 2\}$

$A_3 = \{1, 2\}$
$A_4 = \{1, 2\}$
$A_5 = \{1, 2\}$
$A_6 = \{1, 2, 3\}$
$A_7 = \{1, 2\}$
$A_8 = \{1, 2, 3\}$
$A_9 = \{1, 2\}$
$A_{10} = \{1, 2\}$

The coded values of the decision attribute are '1', '2' and '3'. So there is a set for the decision attribute.

$D = \{1, 2, 3\}$

The plants are scored on each attribute based on presence of a characteristic. For example, for A1=2 (ownership is state-owned), the plants P_2, P_8 and P_{10} have the same score, namely are all state-owned systems. Hence the set *U* could be classified into subsets based on the scored value of each attribute, namely

$$U / A_1 = \{\{P_1, P_3, P_4, P_5, P_6, P_7, P_9\}, \{P_2, P_8, P_{10}\}\}$$
$$U / A_2 = \{\{P_1, P_5, P_8, P_9, P_{10}\}, \{P_2, P_3, P_4, P_6, P_7\}\}$$
$$U / A_3 = \{\{P_1, P_4, P_8, P_9, P_{10}\}, \{P_2, P_3, P_5, P_6, P_7\}\}$$
$$U / A_4 = \{\{P_1, P_2, P_3, P_4, P_5, P_6, P_7\}, \{P_8, P_9, P_{10}\}\}$$
$$U / A_5 = \{\{P_1, P_2, P_3, P_4, P_5, P_7, P_8, P_9, P_{10}\}, \{P_6\}\}$$
$$U / A_6 = \{\{P_1, P_6, P_7, P_9\}, \{P_4, P_5, P_8, P_{10}\}, \{P_2, P_3\}\}$$
$$U / A_7 = \{\{P_3, P_4, P_6, P_7, P_{10}\}, \{P_1, P_2, P_5, P_8, P_9\}\}$$
$$U / A_8 = \{\{P_5, P_9, P_{10}\}, \{P_1, P_3, P_4, P_7, P_8\}, \{P_2, P_6\}\}$$
$$U / A_9 = \{\{P_1, P_2, P_3, P_4, P_5, P_6, P_7, P_9\}, \{P_8, P_{10}\}\}$$

$$U/A_{10}=\{\{P_2,P_3,P_5,P_8\},\{P_1,P_4,P_6,P_7,P_9,P_{10}\}\}$$

Also, the set U could be classified into the subset in terms of the coded value of the decision attribute.

$$U/D=\{\{P_1,P_4,P_8,P_9,P_{10}\},\{P_2,P_3,P_5,P_7\},\{P_6\}\}$$

If we assume: $Y_1=\{P_1,P_4,P_8,P_9,P_{10}\}$; $Y_2=\{P_2,P_3,P_5,P_7\}$; $Y_3=\{P_6\}$

Then $U/D=\{Y_1,Y_2,Y_3\}$

The subset of Y_1 represents the group of plants that have been stopped, Y_2 represents the group of plants that are interrupted, Y_3 represents the group of plants that continue to function. The aim is to find causality among the data sets. The sets U/A_i (i=1, 2, 3…10) may contain the same subset as what the set U/D contains. If we can find the subset Y_j (j=1, 2, 3) of the set U/D in any of the sets U/A_i (i=1, 2, 3…10), the linkages between D and A_i are identified. For example, a set U/A_m (m=1 or 2 or 3… or 10) containing the subset Y_n (n =1 or 2 or 3) means the group of plants in the nth operational status could be characterized by A_m. In other words, the attribute A_m is the critical factor affecting the decision attribute D. The equation identifying these linkages is as follows:

$$(U/A_i) \cap (U/D) = Y_j \quad (i=1, 2, 3\ldots10, j=1, 2, 3)$$

Through data mining we find:

j=1→ i=3, that means U/A_3 contains Y_1.
j=2→ i=3 and i=5, that means U/A_3 and U/A_5 contains Y_2.
j=3→ i=5, that means U/A_5 contains Y_3.

The above results indicate that only the condition attributes A_3 (Doubts about water quality) and A_5 (Availability of groundwater) are the critical factors affecting the operation of the rainwater harvesting plants in Beijing. Except these two factors, other factors do not have significant links with the operation of rainwater harvesting systems in Beijing.

Accordingly the conditional causal links of an 'if..., then ...' nature could be derived, which is called 'rules' in the rough set method. A rule is a combination of values of condition and decision attributes, specifying the relationships between condition and decision attributes. For example, if the condition attribute A_5 equals 2, then the decision attribute D equals 3. The rules of 'if...then...' of this study are shown in Table 6.4. There are three rules generated, and each rule corresponds to one operational status of the plant.

Table 6.4
Rules for operation of rainwater harvesting plants

Rule number	If	Then
1	A_3=1	D=1
2	A_3=2 and A_5=1	D=2
3	A_5=2	D=3

6.5 Discussion of Results

Table 6.5
Description of the rules and the plants concerned

Rule	If	Then	Plants
1	Rainwater quality is doubted	Operation of plant stops	P_1, P_4, P_8, P_9, P_{10}
2	Rainwater quality is not doubted but groundwater source is sufficient	Operation of plant is interrupted	P_2, P_3, P_5, P_7
3	Groundwater source is not sufficient	Operation of plant is continuous	P_6

Table 6.5 shows the description of the rules and the concerned plants of each rule. Table 6.5 illustrates that rule 1 is supported by five plants (P_1, P_4, P_8, P_9, P_{10}), rule 2 is supported by four plants (P_2, P_3, P_5, P_7) and rule 3 is supported by one plant (P_6). This corresponds to classification of the plants in terms of their operational status.

The first rule states that *if* the plant managers have doubts about the collected rainwater quality, *then* the operation of the plant tends to stop. The people of Beijing believe that rainwater in Beijing contains chemical products because of the serious air pollution in the city. Accordingly farmers doubt that the quality of rainwater in Beijing is suitable for agricultural irrigation. It is surprising that the major reason for the failed operation of rainwater harvesting system is doubt of rainwater quality, a non-technical factor. Generally poor technology level is considered a significant reason leading to operational failure. In this study, whether farmers have doubts about rainwater quality is the crucial point to cause plants to stop operations.

The second rule states that *if* the groundwater sources are sufficient and the plant managers have no doubts about the collected rainwater quality, *then* the operation of the plant tends to be interrupted occasionally. According to interviews, the operation of a plant could stop temporarily for non-technical reasons. The main reason causing this status is low motivation to operate the rainwater harvesting systems. For example, in plant 2, although farmers do not doubt water quality and there are no technical or financial problems, the operation of the plant could not continue. Sufficient groundwater resources and low barriers to obtain groundwater lessen the incentives of plant managers to operate rainwater harvesting. According to the interviews and the literature, the barrier to using groundwater in Beijing is still quite low (Wang and Wang 2005; Yang et al. 2003).

The third rule states that *if* the groundwater sources are scarce, *then* the operation of the plant will continue. Since few plants in the empirical sample are in continuous operation, only one plant (P_6) in the study is a successful case (see Table 6.4). At plant 6, there are no technological problems, the plant manager obtained higher subsidies on the initial investment and it is a privately owned plant. But these factors are not the reasons for the continuous operation of plant 6. The major reason for the continuing operation of the rainwater harvesting system is that there are insufficient groundwater resources at plant 6. The shortage of groundwater resources raises the farmers' incentive to search for and use other irrigation resources, which is often the situation in northwest China.

6.6 Conclusions

The chapter aims to identify the critical drivers of success or failure in the operation of rainwater harvesting plants in Beijing. An artificial intelligence method: rough set analysis is applied in the study. First, a data matrix including information on ten factors and the operational status of ten plants is mapped out through the data coding with '1', '2' or '3'. Second, based on the algorithm of data mining, the links between the condition and decision attributes are extracted from the codified data matrix and the decisive factors affecting the operation of rainwater harvesting plants are identified. Accordingly the conditional causal links of an 'if…, then …' nature (rules) could be derived.

The results show that 'Doubts about the rainwater quality' and 'Availability of groundwater' are two decisive factors determining the operation of the rainwater harvesting systems. Both factors are not technological reasons. The results of rules indicate three points. 1) As long as farmers have doubts about rainwater quality, the operation of the plant cannot continue. 2) Although farmers have no doubts about rainwater quality, they still could not operate the plant continually. The reason is that it is easy to get sufficient irrigation water. 3) If there is no groundwater, the rainwater harvesting systems could operate continually and successfully.

Generally the improvement in technology for rainwater harvesting is considered important, while little attention is paid to non technological factors by researchers. In this study, two critical factors were found: 'doubts about the rainwater quality' and 'availability of groundwater', which are not technological factors but have significant effects on the operation of the rainwater harvesting systems in Beijing.

In order to promote rainwater harvesting for agricultural irrigation in Beijing, the government subsidizes the rainwater harvesting systems and improves the technology to make it suitable to local situations. This may create a sound beginning, while it is not enough to promote rainwater harvesting, a new activity in Beijing, in a sustainable way. The status of operation of a rainwater harvesting plant depends on the confidence and motivation of using rainwater. Improving the confidence and motivation of using rainwater is the key to sustainable and successful operation of rainwater harvesting systems in Beijing. Removing doubts of rainwater quality in people's minds is an important first step. Farmers would use rainwater only if they think rainwater is safe for irrigation. Tests of rainwater quality should be carried out and the results should be shown and explained to the public. Raising barriers to obtain groundwater in areas with sufficient water is the second important step. All kinds of measures such as pricing ground water, prohibiting pumping new wells, and limiting the quantity of ground water pumping may be required. The lessons drawn here may be applied in other Chinese areas or other countries. The evidence may also generate new hypotheses to be statistically tested in a broader review of experiences.

Notes

[5] Chapter 6 is submitted to the journal, *Water Resources Management*, 2010, under review.

7 Conclusions

7.1 Introduction

The previous chapters introduced and explained how to use economics in sustainable urban water management through a study of the water systems in Beijing. This chapter synthesizes and draws conclusions from the study. Section 7.2 summarizes the contents and findings of the research. The conclusions are formulated in Section 7.3 and the contributions of the research are presented in Section 7.4.

7.2 Summary of Contents and Findings

Increasing urban population, climate change and deteriorated environment from pollution lead to the immense challenge to achieve sustainable urban water management. The constraints to sustainable water management sometimes come from non-technical problems. Financial and economic factors can be a significant barrier to the operation of water systems.

The research tries to use economics to analyse water treatment systems, to discover out how and what economic analysis contributes to sustainable water management. The research uses the case of Beijing. Beijing is facing the same difficulties as other mega-cities: increasing population, limited water resources, dry weather and environmental pollution, which causes serious water scarcity in Beijing. To solve the water scarcity in Beijing, two main measures are promoted: wastewater reuse and rainwater harvesting. A financial and economic analysis is carried out on wastewater reuse and rainwater harvesting systems.

The thesis can be divided into three parts: 1) background introductions and objective specification; 2) financial and economic analysis of wastewater reuse; 3) financial and economic analysis of rainwater harvesting.

In Chapter 1 the main factors causing urban water crisis were introduced, these include an increasing urban population, limited water availability, increased climate variability and environmental pollution. Moreover, it presents the realities of water scarcity in Beijing, which include increasing population, continuous droughts and depletion of groundwater stocks. Many new technologies and policies have been implemented to solve the urban water crisis. The thesis uses Beijing as an example, introducing technical measures, reform of governance structures, and water policies. These measures may physically help to release the stress of the water crisis, but they cannot ensure sustainable water management. It frequently occurs that newly constructed water systems do not operate continuously because of non-technical problems such as financial problems. The study of the economics of new systems may be an important part in sustainable urban water management.

The research objective is to carry out an extensive and quantitative economic analysis of sustainable urban water management. The research focuses on wastewater reuse systems and rainwater harvesting systems. Cost benefit analysis is the main economic method in the thesis, so the economic theory of cost benefit analysis is discussed. Furthermore, the essence of the methodological framework presented in the research is a two-part analysis: financial and economic analysis. The financial analysis is from the point of view of project managers while the economic analysis is from the point of view of society. The financial analysis assesses the benefits and costs of the operation of the systems. The economic analysis evaluates the economic, environmental and social effects caused by the urban water systems. In addition to cost benefit analysis, the methods of linear programming and rough set analysis are used in the study.

In Chapters 2 and 3, the wastewater reuse systems in Beijing are studied through cost benefit analysis. According to the evaluation framework of financial and economic analysis (Figure 1.17), the financial and economic feasibility of the decentralized and centralized wastewater reuse systems are determined separately from private and public perspectives. Whether the wastewater reuse plants are profitable or not is evaluated in the financial analysis. After the determination of the financial and economic feasibility of the decentralized and centralized wastewater reuse systems, a comparative analysis is carried out between the decentralized and centralized systems.

In Chapter 2, the results of the financial and economic analysis of the decentralized wastewater reuse systems show that decentralized wastewater reuse systems are economically but not financially feasible. The ratio of benefits to costs in the economic analysis is greater than 1 but the ratio of benefits to costs in the financial analysis is less than 1. From the point of view of government, the decentralized wastewater reuse systems are positive to society, while from the point of view of project managers, the systems cannot operate in the long-term. Further analysis shows that low rate of reclaimed water is one important reason leading to decentralized systems not being financially feasible. The current rate of reclaimed water is fixed by the government, which does not reflect the real cost of reclaiming grey water.

In Chapter 3, the financial and economic analysis of the centralized wastewater reuse systems is carried out. The results show centralized wastewater reuse systems are both economically and financially feasible, because the ratios of benefits to costs in the financial and economic analysis are both greater than 1. There are no financial problems in the centralized systems. From both public and private perspectives, centralized wastewater reuse systems deserve to be promoted.

Then there is a comparative analysis between the decentralized and centralized systems in Chapter 3. The comparative analysis is carried out from five aspects: 1) financial and economic feasibility; 2) environmental and social effects; 3) initial investment; 4) O&M cost; 5) cost recovery. The comparative analysis shows that from the perspective of financial feasibility, the centralized wastewater reuse systems are more competitive than the decentralized systems. But the centralized and decentralized wastewater reuse systems are both economically feasible.

The comparative analysis also shows that decentralized and centralized wastewater reuse systems lead to different effects for environments and society. Decentralized systems are constructed on site, so a large amount of capital for pipe construction for the distribution of reclaimed water is saved. Because the decentralized plants are close to the users, noise and malodors generated during the wastewater treatment process could negatively affect surrounding users. However, the environmental benefits caused by decentralized plants are quantitatively larger than the environmental cost. In contrast, the centralized wastewater reuse systems are usually built in suburban areas, which are far from the area having high population density, so the effects of noise and malodors for users are small. Due to high-energy consumption for water treatment, centralized water treatment plants lead to a high carbon dioxide emission causing environmental cost. Despite that centralized systems cause negative environmental effects, the monetary values of the environmental benefits are larger than the values of the environmental cost. Moreover, regarding the social effects, the centralized wastewater reuse systems benefit improved employment in the region while the decentralized systems help to improve social awareness concerning water saving. But building a centralized system requires demolition of existing buildings and relocation of people, which leads to a large social cost. Moreover, the decentralized systems can cause the social cost of health risks.

Furthermore, the comparative analysis indicates that the initial investments of the two systems are very different. The investments in plants and pipes construction of centralized wastewater reuse systems are high. On the contrary, investments in a decentralized wastewater reuse system are relatively low. In some cases, the cost of pipe construction of a centralized wastewater reuse plant could pay for about 25 decentralized plants. Additionally, building a centralized plant requires additional costs for demolition and relocation of buildings and residences, which are not included in investments of decentralized systems. Although there is a big difference in the initial investments between the two systems, the largest expense in the O&M budget is similar: energy cost.

Because of the scale economy, the unit O&M cost of centralized wastewater reuse systems is lower than that of the decentralized systems. In the centralized water systems, the rate of reclaimed water is higher than the unit O&M cost while in the decentralized systems, the rate of reclaimed water is lower than the unit O&M cost. Hence the rate of O&M cost recovery of the centralized systems is much higher than the decentralized systems.

In Chapters 4, 5 and 6, the rainwater harvesting plants are studied through the methods of cost benefit analysis, linear programming and rough set analysis. Groundwater is the major and traditional water resource, and rainwater is the alternative new water resource.

In Chapter 4, the economic implications of using rainwater and groundwater are studied and compared. In the economic analysis, the economic, environmental and social effects caused by rainwater harvesting plants are identified and calculated by cost benefit analysis. In financial analysis, the net present values of using groundwater and rainwater are compared to determine if rainwater harvesting is financially feasible.

The study finds that the rainwater harvesting systems are economically feasible. Rainwater harvesting systems help to save water and energy for pumping, raise social awareness and improve employment. However they consume resources for construction and may cause risks for agriculture. The values of the benefits caused by rainwater harvesting systems are higher than the values of the cost. This means rainwater harvesting systems have positive effects for society. From the point of view of the government, rainwater harvesting systems are worth being promoted. However, the results of the financial analysis indicate that the financial feasibility of rainwater harvesting systems depends on the charge for groundwater and on the size of the rainwater harvesting systems. If groundwater is not charged, the rainwater harvesting systems are not financially feasible. If groundwater is charged at 2 Yuan/m^3, only large size systems are financially feasible, while small and medium size systems are not financially feasible.

If groundwater is charged at a high level, all scales of rainwater harvesting systems can become financially feasible. However, a higher groundwater charge discourages farming, as groundwater is the main source of water for agricultural irrigation. Hence a study of the relation between the charge for groundwater and the consumption of rainwater is carried out in Chapter 4 to discover how to increase rainwater consumption by charging for groundwater; while not discouraging farming.

There are two parts in Chapter 4: theoretical analysis and application of linear programming to this problem. The theoretical analysis studies the relationship between the cost of groundwater and the consumption of rainwater by analysing the elasticity of groundwater demand graphically. If the cost of groundwater is lower than the elasticity threshold, farmers lack incentives to use rainwater. If the cost of groundwater is higher than the threshold, the rainwater consumption increases. The theoretical analysis also shows that the elasticity threshold of groundwater can move down following a change in the characteristics of rainwater harvesting systems. The linear programming analysis indicates that increasing subsidies and enlarging the size of rainwater harvesting systems can decrease the elasticity threshold of groundwater. In that case, low cost groundwater could still increase consumption of rainwater. It illustrates that a realistic groundwater charge can be determined by taking into account the scale of rainwater harvesting and the subsidies on the initial investment for rainwater harvesting systems. Enlarging the scale of rainwater harvesting systems and increasing subsidies could effectively decrease the O&M cost of rainwater harvesting systems, which can help to achieve the balance of effectively improving the financial feasibility of rainwater harvesting while keeping an eye on farmers' benefits.

Both results of Chapters 4 and 5 indicate that enlarging the size and increasing subsidies are two important factors to improve the performance of rainwater harvesting systems. However, according to interviews with the plant manager, the operations of many rainwater harvesting plants are unsuccessful despite using scale and subsidy advantages. Hence a study to find the decisive factors that significantly affect the operation of the rainwater harvesting plant is carried out in Chapter 6. Based on the literature and interviews with rainwater harvesting plant managers, ten impact factors, including technological and non-technological factors were chosen. It is assumed that the operational status of a rainwater harvesting system can be affected by these ten factors. The methodology of rough set analysis is used in Chapter 6 to identify what are the decisive factors determining the operational success or failure of rainwater harvesting systems.

The study identifying the decisive factors affecting the operation of rainwater harvesting systems concludes that the factors of 'Doubts about the rainwater quality' and 'Availability of groundwater' critically affect the performance of rainwater harvesting systems. This implies that as long as farmers have doubts of rainwater quality, they will not want to use rainwater for agricultural irrigation. If the groundwater is sufficient for farmers and a rainwater harvesting plant is available, farmers would prefer to use groundwater even when they have no doubts about rainwater quality. If there is a groundwater shortage, operation of the rainwater harvesting plant is successful. Therefore removing farmers' doubts of rainwater quality and increasing the barrier of using groundwater could help to solve the poor performance of rainwater harvesting decisively. These two decisive factors are non-technological factors but have significant effects on the operation of rainwater harvesting systems.

7.3 Conclusions

The research shows that alternative water treatment systems are economically feasible. In Beijing, decentralized wastewater reuse systems save the cost of constructing pipes, increase water availability and raise social awareness about water saving despite that they bring noise pollution and health risks. Rainwater harvesting systems help to save water and energy, increase farmers' income and raise social awareness although they lead to agricultural risks and consume natural resources for construction. The positive effects on society from these alternative water systems are greater than the negative effects. So the alternative systems are economically feasible, which implies the alternative water systems deserve to be promoted. However, financial analysis shows that the alternative water systems are not financially feasible. In the decentralized wastewater reuse systems, financial benefits cannot recover the financial cost. In rainwater harvesting systems, using rainwater is financially less attractive than using groundwater. Given this situation, the managers of these plants prefer not to operate the alternative systems continuously

However, the research shows that the traditional water treatment systems are both economically and financially feasible. The values of positive effects on society caused by the traditional water systems are greater than the values of negative effects. It is profitable to invest in traditional water systems.

Comparing economic and financial feasibility between traditional water systems and alternative systems, traditional water systems are currently better than the alternative systems are. This means that the new water systems are not viable alternatives to traditional water systems because the new systems are not financially feasible yet.

The research finds that there are three main reasons why alternative water systems are not financially feasible: low revenues, high unit O&M cost and low subsidies. First, low revenues. Because the charge for reclaimed water is minimal, the revenue obtained from wastewater reuse systems is low. Second, high unit O&M cost. The new water systems usually are small scale so the unit O&M cost for the new water systems is relatively higher than traditional systems. Enlarging the scale of new systems could benefit to decrease the unit O&M cost. Third, subsidies are insufficient. Studies of rainwater harvesting plants show that increasing subsidies can improve financial feasibility of the water systems.

In addition to the financial problems, the research finds some social problems have negative influence on operation of alternative water systems. For instance, farmers doubt the rainwater quality and there is sufficient groundwater for agricultural irrigation. So, farmers have no incentive to use rainwater harvesting systems, which leads to limited or closed rainwater harvesting plants.

Currently, it is rare that there are policies or regulations that aim to solve the financial and social problems in the new water systems or to improve the systems' management. For example, there is no policy on raising the rate of reclaimed water, no policy on increasing the barriers for using groundwater, and no policy of efficiently providing subsidies. The existing measurements and polices mostly emphasize building alternative water treatment systems and improving the technology of water treatment. A large number of alternative water systems were constructed, and are expected to supplement water supply in the city. But, in reality, the objective of supplementing water supply has not been achieved because the rate of utilization of alternative water treatment systems in Beijing is very low (Table 1.2). The financial and social problems are hindering operation of these alternative water systems.

Moreover, the current governance structure does not really facilitate operation of the new water systems because some important stakeholders do not function as expected. For example, the Beijing Water Saving Office is an important stakeholder involved in management of new water systems, and is supposed to provide assistance and supervision to the new systems related to water saving in Beijing. Even though the officials at the Beijing Water Saving Office have known that the new water systems have some financial problems and do not operate continually, they did not deal with these problems.

Therefore more effective policies to solve financial and social problems in the new constructed water systems are required urgently. The strategy of urban water management should be changed from physical measurement of the performance to assuring sustainable operation of the systems.

7.4 Contributions of the Research

The thesis demonstrates, through the case of Beijing how to use economics in managing urban water systems. The thesis is the first to carry out an integrated and quantitative analysis of the economic, environmental and social effects caused by alternative water systems. This thesis evaluates the profitability of investment in and the effects on environment and society of alternative water systems. The economic, environmental and social effects are all determined by monetary values, which is rare in the existing literature.

The thesis shows that economics could benefit to identify the non-technical problems in water systems and can help decision makers to make choices consistent with the long-term well being of the community. Three practical contributions of the research are:

1) The use of economics to identify and quantify the effects of water treatment systems on economics, environment and society. In conventional water management, the externalities of a project used to be neglected so that the project causes negative environmental and social effects. This seriously hinders the operation of a water treatment system in the long-term. So it is very important to know what and how much influence the new activity has on environment and society. Economic study tries to evaluate the effects through quantitative methods and quantify the effects by monetary values. Although the calculated monetary values do not present the real values, it provides a measurement for further comparative study.

2) The use of economics to determine the factors that significantly hinder the plants operation in the long-term. Nowadays water management has developed to a stage that the technology is advanced enough to deal with many physical problems. However, it still requires study and finding solutions to the non-technical problems. The problems of urban water management are not only physical but also financial and social. Through economic study, the non-technical factors that prevent water systems from sustainable development can be identified.

3) The use of economic tools to learn the advantages and disadvantages of different water systems from the point of view of economics. Each system has its own advantages and disadvantages. For instance, centralized wastewater reuse systems have sound cost recovery mecha-

nisms while they require high initial investment. Stakeholders have to make the decisions in terms of their requirements and preferences. Learning and comparing the advantages and disadvantages within different systems could help decision makers to choose the most suitable system efficiently. Braden and Van Ierland (1999) also think that with the help of economics, one could choose the optimal solutions related to different institutional and technological options.

The theoretical contribution of the research is that it proves the importance of considering the viewpoints of different stakeholders in the cost benefit analysis. Researchers suggest that it is important to do cost benefit analysis of a project from different stakeholder perspectives (Campbell and Brown 2003, 2005; Dahmen 2000; Gittinger 1982). Different stakeholders may have conflicting goals or conflicting views on how the goals should be achieved. Doing cost benefit analysis from different stakeholder perspectives can provide complete and accurate information for helping decision makers to choose the suitable alternative. This research carries out cost benefit analysis of water systems separately from the public and private perspectives and hence finds that from the public perspective, alternative water systems are feasible while from private perspective, the systems are not yet feasible.

References

Abeysuriya, K., C. Mitchell and J. Willetts (2005) 'Cost Recovery for Urban Sanitation in Asian Countries: Insurmountable Barrier or Opportunity for Sustainability?', Australia New Zealand Society for Ecological Economics Conference, Massey University, New Zealand.

Ahmad, M. (2000) 'Water Pricing and Markets in the Near East: Policy Issues and Options', *Water Policy* 2: 229-42.

Almagro, A.J. (2005) 'Efficient Use of Subsidies in the Financing of Water and Wastewater Investments', *Water Science and Technology: Water Supply* 15: 197-207.

Angelakis, A.N., L. Bontoux and V. Lazarova (2003) 'Challenges and Prospective for Water Recycling and Reuse in EU Countries', *Water Science and Technology* 3: 59-68.

Aramaki, T., M. Galal and K. Hanaki (2006) 'Estimation of Reduced and Increasing Health Risks by Installation of Urban Wastewater Systems', *Water Science and Technology* 53: 247-52.

Arbues, F., M.A. Garcia-Valinas and R. Martinez-Espineria (2003) 'Estimation of Residential Water Demand: A State of the Art Review', *Journal of Socio-Economics* 32: 81-102.

Arrow, K.J., M.L. Cropper, G.C. Eads, R.W. Hahn, L.B. Lave, R.G. Noll, P.R. Portney, M. Russell, R. Schmalensee, V.K. Smith and R.N. Stavins (1996) 'Is There a Role for Benefit Cost Analysis in Environmental, Health, and Safety Regulation?', *Science* 272: 221-2.

Asano, T. (2001) 'Water from (Waste) Water—the Dependable Water Resource', 11th Stockholm Water Symposium, Stockholm, Sweden.

Asano, T. (2005) 'Urban Water Recycling', *Water Science and Technology* 51: 83-9.

Asano, T. and A.D. Levine (1996) 'Wastewater Reclamation, Recycling and Reuse: Past, Present, and Future', *Water Science and Technology* 33: 1-14.

Ashley, R.M., N. Souter, D. Butler, J. Davies, J. Dunkerley, and S. Hendry (1999) 'Assessment of the Sustainability of Alternatives for the Disposal of Domestic Sanitary Waste', *Water Science and Technology* 39: 251-8.

Bao, C. and C. Fang (2007) 'Water Resource Constraint Force on Urbanization in Water Deficient Regions: A Case Study of the Hexi Ccorridor, Arid Area of NW China', *Ecological Economics* 62: 508-17.

Barbagallo, S., S. Consoli, N. Pappalardo, S. Greco, and S.M. Zimbone (2006) 'Discovering Reservoir Operating Rules by a Rough Set Approach', *Water Resources Management* 20: 19-36.

Baycan-Levent, T. and P. Nijkamp (2009) Planning and Management of Urban Green Spaces in Europe: Comparative Analysis', *Journal of Urban Planning and Development* 135(1):1-12.

Becker, N. (1994) 'Value of Moving from Central Planning to a Market System: Lessons from the Israeli Water Sector', *Agricultural Economics* 12: 11-21.

Beijing Water Authority (1986-2009) 'Beijing Water Resources Bulletin', *The Working Papers of Beijing Water Authority*.

Beijing Water Authority (2002a) 'Analysis of Wastewater Reuse in Beijing', *The Working Papers of Beijing Water Authority*.

Beijing Water Authority (2002b) 'Pricing the Reclaimed Water in Beijing', *The Working Papers of Beijing Water Authority*.

Beijing Water Authority (2004) 'The Research of the Gao Bei Dian Wastewater Reclamation System', *The Working Papers of Beijing Water Authority*.

Berbel, J. and J.A. Gomez-Limon (2000) 'The Impact of Water-Pricing Policy in Spain: An Analysis of Three Irrigated Areas', *Agricultural Water Management* 43: 219-38.

Berkoff, J. (2003) 'China: the South-North Water Transfer Project, is it Justified?', *Water Policy* 5: 1-28.

Berthelot, P.B. and C.A. Robertson (1990) 'A Comparative Study of the Financial and Economic Viability of Drip and Overhead Irrigation of Sugarcane in Mauritius', *Agricultural Water Management* 17: 307-15.

Birol, E., K. Karousakis and P. Koundouri (2005) 'Using a Choice Experiment to Estimate Non-use Values: The Case of Cheimaditida Wetland', *Water Science and Technology: Water Supply* 5: 125-85.

Boers, T.M. and J. Ben-Asher (1982) 'A Review of Rainwater Harvesting', *Agricultural Water Management* 5: 145-58.

Borboudaki, K.E., N.V. Paranychianakis and K.P. Tsagarakis (2005) 'Integrated Wastewater Management Reporting at Tourist Areas for Recycling Purposes, Including the Case Study of Hersonissos, Greece', *Environmental Management* 36: 610-23.

Braden, J.B. and E.C. Van Ierland (1999) 'Balancing: The Economic Approach to Sustainable Water Management', *Water Science and Technology* 39: 17-23.

Brent, R.J. (1996) *Applied Cost Benefit Analysis*. US: Edward Elgar.

Brown, L.R. (2004) *China's Shrinking Grain Harvest: How its Growing Grain Imports will Affect World Food Prices*. Earth Policy Institute.

Camagni, R., M.C. Gibelli and P. Rigamonti (2002) 'Urban Mobility and Urban Form: The Social and Environmental Costs of Different Patterns of Urban Expansion', *Ecological Economics* 40: 199-216.

Campbell, H. and R. Brown (2003) *Benefit-Cost Analysis: Financial and Economic Appraisal Using Spreadsheets*. Melbourne: Cambridge University Press.

Campbell, H. and R. Brown (2005) 'A Multiple Account Framework for Cost Benefit Analysis', *Evaluation and Program Planning* 28: 23-32.

Chen, D., J. Chen and Z. Luo (2006) 'Evaluation Method of Natural Water Resources Based on Emergy Theory and its Application', *Shui Li Xue Bao (in Chinese)* 37: 1188-92.

Christova-Boal, D., R.E. Eden and S. McFarlane (1996) 'An Investigation into Greywater Reuse for Urban Residential Properties', *Desalination* 106: 391-7.

Chu, J., J. Chen, C. Wang and P. Fu (2004) 'Wastewater Reuse Potential Analysis: Implications for China's Water Resources Management', *Water Research* 38(11): 2746-56.

Dahmen, E.R. (2000) 'Financial and Economic Analysis', UNESCO-IHE Institute for Water Education (lecture notes HH334/00/1).

Dahowski, R.T., X. Li, C.L. Davidson, N. Wei, J.J. Dooley and R.H. Gentile (2009) 'A Preliminary Cost Curve Assessment of Carbon Dioxide Capture and Storage Potential in China', *Energy Procedia* 1(1): 2849-56.

Deng, F. and W. Chen (2003) 'Rainwater Reuse Plan for Residential Districts in Nanjing', *Urban Environment and Urban Ecology (in Chinese)* 16: 104-6.

DPP (2001) *The Water Planning in Beijing*. Chinese Department of Planning and Program.

Fane, S.A., N.J. Ashbolt and S.B. White (2002) 'Decentralized Urban Water Reuse: The Implications of System Scale for Cost and Pathogen Risk', *Water Science and Technology* 46: 281-8.

Fang, L. and E.A. Nuppenau (2004) 'A Spatial Model (SWAM) for Water Efficiency and Irrigation Technology Choices using GAMS - A Case Study from Northwestern China', Seventh Annual Conference on Global Economic Analysis, Washington DC.

Feng, Y., D. He and K. Beth (2006) 'Water Resources Administration Institutions in China', *Water Policy* 8: 291-301.

Friedler, E. and M. Hadari (2006) 'Economic Feasibility of Sn-site Greywater Reuse in Multi-story Buildings', *Desalination* 190: 221-34.

Ganderton, P.T. (2005) '"Benefit Cost Analysis" Of Disaster, Mitigation: Application as a Policy and Decision Making Tool', *Mitigation and Adaptation Strategies for Global Change* 10: 445-65.

Genius, M., M. Manioudaki, E. Mokas, E. Pantagakis, D. Tampakakis and K.P. Tsagarakis (2005) 'Estimation of Willingness to Pay for Wastewater Treatment', *Water Science and Technology: Water Supply* 5: 105-13.

Gintis, H. (2000) 'Beyond Homo Economicus: Evidence from Experimental Economics', *Ecological Economics* 35: 311-22.

Gittinger, J.P. (1982) *Economic Analysis of Agricultural Projects* (2nd edn). Baltimore, MD: The Johns Hopkins University Press.

Goel, A.K. and R. Kumar (2004) 'Economic Analysis of Water Harvesting in a Mountainous Watershed in India', *Agricultural Water Management* 71: 257-66.

Gratziou, M., S. Ekonomou and M. Tsalkatidou (2005) 'Cost Analysis and Evaluation of Urban Sewage Processing Units', *Water Sciences and Technology: Water Supply* 5: 155-62.

Hanley, N and Spash, C L, 1993. *Cost-Benefit Analysis and the Environment.* UK: Edward Elgar.

Hatibu, N., K. Mutabazi, E.M. Senkondo and A.S.K. Msangi (2006) 'Economics of Rainwater Harvesting for Crop Enterprises in Semi-arid Areas of East Africa', *Agricultural Water Management* 80: 74-86.

Hauger, M.B., W. Rauch, J.J. Linde and P.S. Mikkelsen (2002) 'Cost Benefit Risk: A Concept for Management of Integrated Urban Wastewater Systems?', *Water Science and Technology* 45: 185-93.

He, X-F, H. Cao and F-M Li (2007) 'Econometric Analysis of the Determinants of Adoption of Rainwater Harvesting and Supplementary Irrigation Technology (Rhsit) in the Semiarid Loess Plateau of China', *Agricultural Water Management* 89: 243-50.

Helmreich, B. and H. Horn (2009) 'Opportunities in Rainwater Harvesting', *Desalination* 248(1-3): 118-24.

Hernandez, F., A. Urkiaga, D.L. Fuentes, B. Bis, E. Chiru, B. Balazs and T. Wintgens (2006) 'Feasibility Studies for Water Reuse Projects: An Economical Approach', *Desalination* 187: 253-61.

Hou, E. (2000) 'Briefing Paper on Water Governance Structure in Beijing, PRC', Accessed 9 June 2009

<http://www.chs.ubc.ca/china/water%20governance.pdf>.

Jenerette, G.D. and L. Larsen (2006) 'A Global Perspective on Changing Sustainable Urban Water Supplies', *Global and Planetary Change* 50: 202-11.

Jia, H., R. Guo, K. Xin and J. Wang (2005) 'Research on Wastewater Reuse Planning in Beijing Central Region', *Water Science and Technology* 51: 195-202.

Kang, J. and K. Meng (1994) 'The Effects of Air Pollution on the Agricultural Production in Fenxi City, China', *Environmental Protection (in Chinese)* 12: 40-1.

Kim, H. (1995) 'Marginal Cost and Second-best Pricing for Water Services', *Review of Industrial Organization* 10: 323-38.

Koutsoyiannis, A . (1979) *Modern Microeconomics.* USA: Macmillan Education.

Kuang, Y. and D. Sun (1998) 'Degradation of Natural Resources by Environmental Pollution - The Economic Loss Caused by Atmospheric

Growth Variation in Huizhou with that in Zhaoqing', *Geochimica (in Chinese)* 27: 373-84.

Larsen, T.A. and W. Gujer (1997) 'The Concept of Sustainable Urban Water Management', *Water Science and Technology* 35: 3-10.

Lazarova, V., B. Levine, J. Sack, G. Cirelli, P. Jeffrey, H. Muntau, M. Salgot and F. Brissaud (2001) 'Role of Water Reuse for Enhancing Integrated Water Management in Europe and Mediterranean Countries Water Science & Technology', *Water Science and Technology* 43: 25-33.

Lee, S. (2006) 'China's Water Policy Challenges', Discussion Paper 13, China Policy Institute, University of Nottingham.

Lens, P., G. Zeeman and G. Lettinga (2001) *Decentralized Sanitation and Reuse: Concept, System and Implementation.* Cornwall, UK: IWA Publishing.

Li, F., S. Cook, G.T. Geballe and W.R. Burch (2000) 'Rainwater Harvesting Agriculture: An Integrated System for Water Management on Rainfed Land in China's Semiarid Areas', *Ambio,* 29(8): 477-83.

Li, H. (2003) 'A Study on the Relation between Economic Growth and Employment Elasticity', *Journal of Finance and Economics (in Chinese)* 29.

Lipsey, M.W. and D.B. Wilson (2001) *Practical Meta-analysis.* London: SAGE Publications.

Liu, C. (1998) 'Environmental Issues and the South-North Water Transfer Scheme', *The China Quarterly*, 156: 899-910.

Liu, F.X. (1999) 'The Study on the Monetary Cost of Noise Pollution in Dalian City', *Liaoning Urban Environmental Technology (in Chinese)* 19: 27-9.

Liu, X. and X. Chen (2003) 'The Evaluation Model of Shadow Price of Water', *Technological Progress of Water and Energy (in Chinese)* 23.

Luo, Y. (2007) *The Social Cost of Residential Resettlement in China.* Master thesis (in Chinese), Si Chuan University.

Malpezzi, S. (1999) 'Estimates of the Measurement and Determinants of Urban Sprawl in US Metropolitan Areas', Working Papers, The Center for Urban Land Economics Research, The University of Wisconsin.

Mao, J.C.T. (1966) 'Efficiency in Public Urban Renewal Expenditures through Benefit-Cost Analysis', *Journal of the American Planning Association* 32: 95-107.

Maurer, M. (2009) 'Specific Net Present Value: An Improved Method for Assessing Modularisation Costs in Water Services with Growing Demand', *Water Research* 43: 2121-30.

Maurer, M., D. Rothenberger and T.A. Larsen (2006) 'Decentralized Wastewater Treatment Technologies from a National Point of View: At What Cost are They Competitive?', *Water Science and Technology: Water Supply* 5: 145-54.

Mishan, E.J. (1988) *Cost-benefit Analysis*. London: Unwin Hyman.

Mushtaq, S., D. Dawe and M. Hafeez (2007) 'Economic Evaluation of Small Multi-purpose Ponds in the Zhanghe Irrigation System, China', *Agricultural Water Management* 91: 61-70.

Nigam, A. and S. Rasheed (eds) (1998) 'Financing of Fresh Water for All: A Right Based Approach', Evaluation, Policy and Planning, Department of Economic and Social Affairs, United Nations.

Nijkamp, P., P. Rietveld and L. Spierdijk (2000) 'A Meta-analytic Comparison of Determinants of Public Transport Use: Methodology and Application', *Environment and Planning B: Planning and Design* 27: 893-903.

Nijkamp, P., M. Van der Burch and G. Vindigni (2002) 'A Comparative Institutional Evaluation of Public-Private Partnerships in Dutch Urban Land-use and Revitalisation Projects', *Urban Studies* 39: 1865-80.

Norton, J.W. (2009) 'Decentralized Systems', *Water Environment Research* 81: 1440-50.

Nurizzo, C., L. Bonomo and F. Malpei (2001) 'Some Economic Considerations on Wastewater Reclamation for Irrigation, with Reference to the Italian Situation', *Water Science and Technology* 43: 75-81.

OECD (2007) 'Unsafe Water, Sanitation and Hygiene: Associated Health Impact and the Costs and Benefits of Policy Interventions at the Global Level', Working Party on National Environmental Policies.

Ogoshi, M., Y. Suzuki and T. Asano (2001) 'Water Reuse in Japan', *Water Science and Technology* 43: 17-23.

Osman, S., A.H. Nawawi and J. Abdullah (2008) 'Urban Sprawl and its Financial Cost: A Conceptual Framework', *Asian Social Science* 4: 39-50.

Ottoson, J. and T.A. Stenström (2003) 'Faecal Contamination of Greywater and Associated Microbial Risks', *Water Research* 37: 645-55.

Oweis, T. and A. Hachum (2006) 'Water Harvesting and Supplemental Irrigation for Improved Water Productivity of Dry Farming Systems in West Asia and North Africa', *Agricultural Water Management* 80(1-3):57-73.

Pandey, S. (1991) 'The Economics of Water Harvesting and Supplementary Irrigation in the Semi-arid Tropics of India', *Agricultural Systems* 36: 207-20.

Partzsch, L. (2009) 'Smart Regulation for Water Innovation - The Case of Decentralized Rainwater Technology', *Journal of Cleaner Production* 17: 985-91.

Pawlak, Z. (1982) 'Rough Sets', *International Journal of Computer and Information Sciences* 11: 341-56.

Pearce, D.W. and C.A. Nash (eds) (1981) *The Social Appraisal of Projects: A Text in Cost-benefit Analysis.* Basingstoke: Macmillan Education Ltd.

Pearce, W.B. (1989) *Communication and the Human Condition.* Illinois: Southern Illinois University Press.

Postel, S.L., G.C. Daily and P.R. Ehrlich (1996) 'Human Appropriation of Renewable Fresh Water', *Science* 271: 785-8.

Prest, A.R. and R. Turvey (1968) 'Cost Benefit Analysis: A Survey', *Economic Journal* 75: 683-735.

Psychoudakis, A., A. Ragkos and M. Seferlis (2005) 'An Assessment of Wetland Management Scenarios: The Case of Zazari-Cheimaditida (Greece)', *Water Science and Technology: Water Supply* 5: 115-24.

Rawski, T.G. (1979) 'Economic Growth and Employment in China', *World Development* 7: 767-82.

Rees, J.A. (2006) 'Urban Water and Sanitation Services: An IWRM Approach', Tec Background Papers.

Renzetti, S. (1999) 'Evaluating the Welfare Effects of Reforming Municipal Water Prices', *Journal of Environmental Economics and Management* 22: 147-63.

Renzetti, S. and J. Kushner (2000) 'The Under Pricing of Water Supply and Sewage Treatment'. Ontario Water Conference: Challenge and Solutions, Toronto.

Rogers, P., R. Silva and R. Bhatia (2002) 'Water is an Economic Good: How to Use Prices to Promote Equity, Efficiency, and Sustainability', *Water Policy* 4: 1-17.

Salman, A.Z. and E. Al-Karablieh (2004) 'Measuring the Willingness of Farmers to Pay for Groundwater in the Highland Areas of Jordan', *Agricultural Water Management* 68(1): 61-76.

Scheierling, S.M., J.B. Loomis and R.A. Young (2004) 'Irrigation Water Demand: A Meta Analysis of Price Elasticities', The American Agricultural Economics Association Annual Meeting, Denver.

Seckler, D., U. Amarasinghe, D. Molden, R. de Silva and R. Barker (1998) 'World Water Demand and Supply, 1990 to 2025: Scenarios and Issues'. Research Report 19, International Water Management Institute <http://www.iwmi.cgiar.org>.

Seppala, O.T. and T.S. Katko (2003) 'Appropriate Pricing and Cost Recovery in Water Services', *Journal of Water Supply: Research and Technology* 52: 225-36.

Singh, M.R., V. Upadgyay and A.K. Mittal (2005) 'Urban Water Tariff Structure and Cost Recovery Opportunities in India', *Water Science and Technology* 52: 43-51.

Skeer, J. and Y. Wang (2005) 'Carbon Charges and Natural Gas Use in China', *Energy Policy* 34: 2251-62.

Slowinski, R. (1991) *Intellligent Decison Support: Handbook of Applications and Advances of Rough Set Theory*. Norwell, MA: Kluwer Academic Publishers.

Song, J., M. Han, T. Kim and J. Song (2009) 'Rainwater Harvesting as a Sustainable Water Supply Option in Banda Aceh', *Desalination* 248(1-3):233-40.

Song, Q., Q. Zuo and F. Yang (2004) 'Water Problems Brought out by Citifying Construction and Measures', *Journal of Water Resource and Water Engineering (in Chinese)* 15: 56-8.

Tian, J., H. Xiao and S. Yao (2007) 'Analysis of Rainwater Harvesting Models in Beijing (in Chinese)', *Water Saving and Irrigation* 3: 60-1.

Tian, Y., F. Li and P. Liu (2002) 'Economic Analysis of Rainwater Harvesting and Irrigation Methods, with an Example from China', *Agricultural Water Management* 60: 217-26.

Tol, R.S.J. (2005) 'The Marginal Damage Costs of Carbon Dioxide Emissions: An Assessment of the Uncertainties', *Energy Policy* 33: 2064-74.

Tol, R.S.J. (2008) 'The Social Cost of Carbon: Trends, Outliers and Catastrophes', *Economics: The Open-Access, Open-Assessment E-Journal* 2(25).

Tsagarakis, K.P. (2005) 'New Directions in Water Economics, Finance and Statistics', *Water Science and Technology: Water Supply* 5: 1-15.

Tsagarakis, K.P., D.D. Mara, N.J. Horan and A.N. Angelakis (2000) 'Small Municipal Wastewater Treatment Plants in Greece', *Water Science and Technology* 41: 41-8.

Tziakis, I., I. Pachiadakis, M. Moraitakis, K. Xideas, G. Theologis and K.P. Tsagarakis (2008) 'Valuing Benefits from Wastewater Treatment and Reuse using Contingent Valuation Methodology', *Desalination* 237: 117-25.

UN (2004) *World Population Prospects: The 2004 Revision.* New York: UN.

Varela-Ortega, C., J.M. Sumpsi, A. Garrido, M. Blanco and E. Iglesias (1998) 'Water Pricing Policies, Public Decision Making and Farmers' Response: Implications for Water Policy', *Agricultural Economics* 19: 193-202.

Wang, B. (2007) 'Wastewater Reclamation and Reuse in Chinese Cities (in Chinese)'. Accessed 9 June 2009 <www. chinacitywater.org>.

Wang, K., C. Li and Z. Wang (2007) 'Demonstration of Rainwater Harvesting and Reuse in Beijing', Beijing Municipal Bureau of Agriculture.

Wang, L. and C. Ma (1999) 'A Study on the Environmental Geology of the Middle Route Project of the South–North Water Transfer', *Engineering Geology* 51:153-65.

Wang, X.C., R. Chen, Q.H. Zhang and K. Li (2008) 'Optimized Plan of Centralized and Decentralized Wastewater Reuse Systems for Housing Development in the Urban Area of Xi' an, China', *Water Science and Technology* 58: 969-75.

Wang, Y. and H. Wang (2005) 'Sustainable Use of Water Resources in Agriculture in Beijing: Problems and Countermeasures', *Water Policy* 7: 345-57.

Webber, M., J. Barnett, B. Finlayson and M. Wang (2008) 'Pricing China's Irrigation Water', *Global Environmental Change* 18: 617-25.

WHO (2004) 'Global Burden of Disease Report'. Accessed 10 June 2009 <www.who.org>.

WHO (2007) 'Estimating the Costs and Health Benefits of Water and Sanitation Improvements at Global Level', *Journal of Water and Health* 05: 467.

Wilderer, P.A. and D. Schreff (2000) 'Decentralized and Centralized Wastewater Management: A Challenge for Technology Developers', *Water Science and Technology* 41: 1-8.

World Bank (2007) 'Analysis on China Northern Cities' Water Management'. Accessed 10 June 2009 <http://www.worldbank.org.cn/Chinese/content/ncwqms_cn.pdf>.

World Bank (2007) 'Cost of Pollution in China: Economic Estimates of Physical Damages'. Retrieved 10 June 2009 <www.worldbank.org/eapenvironment>.

Wu, C., Y. Yue, M. Li and O. Adjei (2004) 'The Rough Set Theory and Applications', *Engineering Computations* 21: 488-511.

Yamagata, H., M. Ogoshi, Y. Suzuki, M. Ozaki and T. Asano (2003) 'On-site Water Recycling Systems in Japan', *Water Science and Technology* 3: 149-54.

Yang, H. and K.C. Abbaspour (2007) 'Analysis of Wastewater Reuse Potential in Beijing', *Desalination* 212: 238-50.

Yang, H. and A.J.B. Zehnder (2001) 'China's Regional Water Scarcity and Implications for Grain Supply and Trade', *Environment and Planning* 33(1): 79-95.

Yang, H., X. Zhang and A.J.B. Zehnder (2003) 'Water Scarcity, Pricing Mechanism and Institutional Reform in Northern China Irrigated Agriculture', *Agricultural Water Management* 61: 143-61.

Zhang, Z. and Z. Xu (2009) 'Rough Set Method to Identify Key Factors Affecting Precipitation in Lhasa', *Stochastic Environmental Research and Risk Assessment* 23: 1181-6.

Zuo, J., C. Liu and H. Zheng (2010) 'Cost - Benefit Analysis for Urban Rainwater Harvesting in Beijing', *Water International* 35: 195-209.

Appendix A

Appendix A.1 Entrance to the Qingzhiyuan plant (studied in Chapter 2)
Source: taken by author

Appendix A.2 Oxygen tank at the Qingzhiyuan plant (studied in Chapter 2)
Source: taken by author

Appendix A.3 Gaobeidian plant (studied in Chapter 3)
Source: Beijing Water Authority

Appendix A.4 Tank at the Gaobeidian plant (studied in Chapter 3)
Source: Beijing Water Authority

Appendix A.5 Jiuxianqiao plant (studied in Chapter 3)
Source: Beijing Water Authority

Appendix A.6 Tank at the Jiuxianqiao plant (studied in Chapter 3)
Source: Beijing Water Authority

Appendix B

Appendix B.1 A green house with the plastic cover (studied in Chapters 4, 5 and 6)
Source: taken by author

Appendix B.2 The ditch in front of the green house (studied in Chapters 4, 5 and 6)

Source: taken by author

Appendix B.3 Storage tank under construction (studied in Chapters 4, 5 and 6)

Source: taken by author

Appendix B.4 The cover of the sediment tank, which is connected with the ditch (studied in Chapters 4, 5 and 6)

Source: taken by author

Appendix B.5 The pipes transferring collected rainwater from storage tank to plants

(studied in Chapters 4, 5 and 6)

Source: taken by author

Curriculum Vitae

In October 2011, Ms Xiao Liang expects to defend her Doctor of Philosophy (PhD) degree at the International Institute of Social Studies (ISS) of Erasmus University Rotterdam, The Netherlands. From 2006 to 2010, Ms Liang conducted her PhD research at UNESCO-IHE Institute for Water Education, Delft, The Netherlands. Her PhD research was carried out within the framework of the European research project SWITCH (Sustainable Urban Water Management Improves Tomorrow's City's Health). SWITCH is supported by the European Commission under the 6th Framework Programme and contributes to the thematic priority area of 'Global Change and Ecosystems' [1.1.6.3] Contract n° 018530-2. Ms Liang obtained her MA degree in Development Economics in 2005 from the Institute of Social Studies, The Netherlands and BA degree in Economics in 2002 from Shanghai Jiao Tong University, China. Ms Liang is the author/co-author of 12 research papers on urban water management and a reviewer for the international journals *Water Resource Management* and *Journal of Environmental Management.* Her research interests include water economics, environmental economics, urban water management and sustainable development and management.

Journal Publications

Liang, X. and Van Dijk, M. P. (2010) 'Financial and Economic Feasibility of Decentralized Wastewater Reuse Systems in Beijing', *Water Science and Technology* 61(8): 1965-73.

Liang, X. and Van Dijk, M. P. (2011) 'Economic and Financial Analysis on Rainwater Harvesting for Agricultural Irrigation in the Rural Area of Beijing', *Resources, Conservation and Recycling*, Accepted.

Liang, X. and Van Dijk, M. P. (2010) 'Decisive Factors Affecting Rainwater Harvesting in Beijing', *Water Resources Management* (submitted).

Liang, X. and Van Dijk, M. P. (2010) 'Groundwater Charge and Rainwater Consumption for Agricultural Water Management in Beijing', *Water* (submitted).

Liang, X. and Van Dijk, M. P. (2010) 'Cost Benefit Analysis of Centralized Wastewater Reuse Systems', *Journal of Benefit-Cost Analysis* (submitted).

Conference Publications

Liang, X. and Van Dijk, M. P. (2009) 'Financial Analysis on Rainwater Reuse for Agricultural Irrigation in Beijing', paper presented at The International Conference in Sustainable Development in Building and Environment, Chongqing, China.

Van Dijk, M. P. and Liang, X. (2009) 'Water Governance in the Water and Sanitation Sector in Beijing, the Capital of China', paper presented at Scientific Meeting on Water Governance in SWITCH Cities, Delft, The Netherlands.

Liang, X. and Van Dijk, M. P. (2009) 'The Use of Economic Science in SWITCH', paper presented at The 4th SWITCH Scientific Meeting, Delft, The Netherlands.

Liang, X. and Van Dijk, M. P. (2009) 'Rainwater Harvesting and Reuse for Agricultural Irrigation in Beijing', presented at The CERES Summer School, Center for International Development Issues, Radboud University Nijmegen, The Netherlands.

Liang, X. and Van Dijk, M. P. (2008) 'Economic and Financial Analysis of Decentralized Water Recycling Systems in Beijing', paper presented at The 3rd SWITCH Scientific Meeting, Belo Horizonte, Brazil.

Liang, X. and Van Dijk, M. P. (2007) 'Economic Implications of More Ecological Urban Water Systems', paper presented at The 2nd SWITCH Scientific Meeting, Tel Aviv, Israel.

Liang, X. and Van Dijk, M. P. (2007) 'Financing and Cost Recovery of Innovations in the Urban Water Cycle in Terms of Different Institutional and Technological Options', paper presented at The 1st SWITCH Scientific Meeting, Birmingham, UK.

Contacts

E-mail: x.liang@yahoo.com TEL: (+852) 53459342

* 9 7 8 0 4 1 5 6 9 1 7 3 4 *

T - #0108 - 230425 - C10 - 240/170/11 - PB - 9780415691734 - Gloss Lamination